168
Structure and Bonding

Series Editor:

D.M.P. Mingos, Oxford, United Kingdom

Editorial Board:

F.A. Armstrong, Oxford, United Kingdom
X. Duan, Beijing, China
L.H. Gade, Heidelberg, Germany
K.R. Poeppelmeier, Evanston, IL, USA
G. Parkin, NewYork, USA
M. Takano, Kyoto, Japan

Aims and Scope

The series *Structure and Bonding* publishes critical reviews on topics of research concerned with chemical structure and bonding. The scope of the series spans the entire Periodic Table and addresses structure and bonding issues associated with all of the elements. It also focuses attention on new and developing areas of modern structural and theoretical chemistry such as nanostructures, molecular electronics, designed molecular solids, surfaces, metal clusters and supramolecular structures. Physical and spectroscopic techniques used to determine, examine and model structures fall within the purview of *Structure and Bonding* to the extent that the focus is on the scientific results obtained and not on specialist information concerning the techniques themselves. Issues associated with the development of bonding models and generalizations that illuminate the reactivity pathways and rates of chemical processes are also relevant

The individual volumes in the series are thematic. The goal of each volume is to give the reader, whether at a university or in industry, a comprehensive overview of an area where new insights are emerging that are of interest to a larger scientific audience. Thus each review within the volume critically surveys one aspect of that topic and places it within the context of the volume as a whole. The most significant developments of the last 5 to 10 years should be presented using selected examples to illustrate the principles discussed. A description of the physical basis of the experimental techniques that have been used to provide the primary data may also be appropriate, if it has not been covered in detail elsewhere. The coverage need not be exhaustive in data, but should rather be conceptual, concentrating on the new principles being developed that will allow the reader, who is not a specialist in the area covered, to understand the data presented. Discussion of possible future research directions in the area is welcomed.

Review articles for the individual volumes are invited by the volume editors.

In references *Structure and Bonding* is abbreviated *Struct Bond* and is cited as a journal.

More information about this series at http://www.springer.com/series/430

Chunming Xu • Quan Shi
Editors

Structure and Modeling of Complex Petroleum Mixtures

With contributions by

Z. Chen · K.H. Chung · S.R. Horton · Z. Hou · J. Ji ·
M.T. Klein · S. Li · Q. Shi · J. Wang · L. Wang · C. Xu ·
Q. Yang · D. Zhai · L. Zhang · L. Zhao · S. Zhao · X. Zhao ·
H. Zheng

Springer

Editors
Chunming Xu
State Key Laboratory of Heavy
  Oil Processing
China University of Petroleum
Beijing, China

Quan Shi
State Key Laboratory of Heavy
  Oil Processing
China University of Petroleum
Beijing, China

ISSN 0081-5993   ISSN 1616-8550 (electronic)
Structure and Bonding
ISBN 978-3-319-32320-6   ISBN 978-3-319-32321-3 (eBook)
DOI 10.1007/978-3-319-32321-3

Library of Congress Control Number: 2016939983

© Springer International Publishing Switzerland 2016
This work is subject to copyright. All rights are reserved by the Publisher, whether the whole or part of the material is concerned, specifically the rights of translation, reprinting, reuse of illustrations, recitation, broadcasting, reproduction on microfilms or in any other physical way, and transmission or information storage and retrieval, electronic adaptation, computer software, or by similar or dissimilar methodology now known or hereafter developed.
The use of general descriptive names, registered names, trademarks, service marks, etc. in this publication does not imply, even in the absence of a specific statement, that such names are exempt from the relevant protective laws and regulations and therefore free for general use.
The publisher, the authors and the editors are safe to assume that the advice and information in this book are believed to be true and accurate at the date of publication. Neither the publisher nor the authors or the editors give a warranty, express or implied, with respect to the material contained herein or for any errors or omissions that may have been made.

Printed on acid-free paper

This Springer imprint is published by Springer Nature
The registered company is Springer International Publishing AG Switzerland

Preface

Petroleum is an extremely complex hydrocarbon mixture, comprised of heteroatoms such as sulfur-, nitrogen-, and oxygen-containing compounds as well as organometallic compounds. Using advanced analytical techniques, millions of petroleum compounds have been identified; however, these just account for a fraction of all species found in petroleum. There is still a way to go before the molecular composition of petroleum, especially for heavy oils, is fully understood.

In the downstream petroleum industry, many process operations are designed using the bulk properties of petroleum, such as boiling point distribution, density, viscosity, and elemental analysis. As a result, many petroleum processes are "black box" operations. Molecular simulation/modeling has been used for the development of petroleum process techniques as well as the catalyst design for petroleum refining.

The terms of "petroleomics," "molecular refining," and "molecular management" have long been proposed by both academia and industry that design and optimize processing at the molecular level. The development of molecular-level process models is crucial for the optimization of unit operations to obtain products that meet environmental specifications and quality requirements. The building of compositional models and process simulations based on molecular composition covering separation and reaction are the two key tasks required to realize this objective.

Each of the five chapters of this volume is the result of contributions by researchers from the State Key Laboratory of Heavy Oil Processing (China) and the University of Delaware. The main content is basic and vital to petroleum chemistry and engineering, such as asphaltenes chemistry, molecular structure of heavy oil, molecular simulation, compositional modeling, and process simulations for petroleum refining.

Contents

Molecular Structure and Association Behavior of Petroleum Asphaltene... 1
Zhentao Chen, Linzhou Zhang, Suoqi Zhao, Quan Shi, and Chunming Xu

Porphyrins in Heavy Petroleums: A Review 39
Xu Zhao, Chunming Xu, and Quan Shi

**Ruthenium Ion-Catalyzed Oxidation for Petroleum Molecule
Structural Features: A Review** 71
Quan Shi, Jiawei Wang, Xibin Zhou, Chunming Xu, Suoqi Zhao,
and Keng H. Chung

**Molecular-Level Composition and Reaction Modeling for Heavy
Petroleum Complex System** 93
Zhen Hou, Linzhou Zhang, Scott R. Horton, Quan Shi, Suoqi Zhao,
Chunming Xu, and Michael T. Klein

Molecular Modeling for Petroleum-Related Applications 121
Liang Zhao, Dong Zhai, Huimin Zheng, Jingjing Ji, Lei Wang, Shiyi Li,
Qing Yang, and Chunming Xu

Index 179

Molecular Structure and Association Behavior of Petroleum Asphaltene

Zhentao Chen, Linzhou Zhang, Suoqi Zhao, Quan Shi, and Chunming Xu

Abstract Asphaltenes, the most polar fraction in crude oil, are critical to all aspects of petroleum utilization. The strong interactions between asphaltenes lead to various levels of aggregation, which is responsible for a variety of transportation and upgrading problems. The structure and aggregation of asphaltene have received worldwide concerns, and a lot of efforts have been made to characterize asphaltene aggregates and related phenomena. The complexity of asphaltene composition makes it difficult to understand the true nature of aggregation. Advanced instruments have been applied to characterize the structure of asphaltenes and its aggregates and also their association behavior. The recent approaches on both analytical measurement and modeling lead to new insights into asphaltene structure and aggregation processes. This has led to new aggregate architecture and aggregation mechanisms. Modeling approaches were also used to predict association energies, aggregate size distribution, and phase behavior. A lot of model compounds have been synthesized or built on the computer to help understand the interaction between molecules. The results challenged the traditional view that heavy petroleum and asphaltene are ultra-large molecules. The asphaltenes are composed of numerous small molecules with strong molecular interactions which form complex nanoaggregates. In this review, we make a brief summary of the recent progress on molecular aggregation of asphaltene and discuss new theories, discoveries, and ongoing debates.

Keywords Aggregation • Asphaltene • Association model • Equation of state (EoS) • Heavy petroleum • Molecular simulation • Oil deposition

---

Z. Chen, L. Zhang (✉), S. Zhao, Q. Shi, and C. Xu
State Key Laboratory of Heavy Oil Processing, China University of Petroleum, Beijing 102249, China
e-mail: lzz@cup.edu.cn

Contents

1 Introduction 2
2 Chemical Structure of Asphaltene Molecule 4
   2.1 Molecular Weight (MW) 4
   2.2 Detail Composition and Structure 7
3 Molecular Interactions and Aggregation Models 8
   3.1 Asphaltene Molecular Interactions 8
   3.2 Structural Model of Asphaltene Aggregation 10
4 Asphaltene Deposition 15
   4.1 Asphaltene Micellization and Aggregation 15
   4.2 Asphaltene Nanoaggregation 16
   4.3 Clustering of Asphaltene Nanoaggregates 17
5 Size Distribution of Asphaltene Monomers and Aggregates 18
   5.1 Size of Asphaltene Monomers 18
   5.2 Size of Asphaltene Aggregates 19
   5.3 Effects on Size of Asphaltenes 22
6 Molecular Simulation 23
   6.1 Atomic Level 23
   6.2 Coarse-Grained Level 25
7 Thermodynamic Models and Oil Compatibility 26
8 Concluding Remarks 29
References 30

1 Introduction

In the petroleum industry, asphaltenes are defined operationally as the part of crude oil that are insoluble in low-carbon-number alkane (i.e., *n*-pentane or *n*-heptane) and soluble in strong solvents (e.g., toluene). The definition of asphaltenes inherently indicates that they are just a solubility class having undefined chemical compositions and molecular structures. The molecular behavior of asphaltenes is highly sensitive to the method used for extracting the asphaltene molecules and the properties of the organic media.

Even though the classic classification method does not give a clear definition, the interest on asphaltene is still strong and will remain so in the future. Compared to the *n*-alkane soluble part of crude oil (named maltene), asphaltenes have attracted more attention. They are responsible for many problems during oil production, transport, and refining due to the strong tendency to aggregate. Change of conditions, such as diluent addition and pressure drop, can trigger asphaltene precipitation, and this leads to wellbore or pipeline fouling and plugging [1, 2], formation of stable water-in-oil emulsions [3, 4], sedimentation during crude oil storage [5], and plugging of catalysts [6]. For exploration and processing of petroleum resource, understanding asphaltene structure and its deposition is critically important.

The research on asphaltene covers several dimensions. Figure 1 illustrates a hierarchy structure from molecule to deposit phase in oil production. Generally, asphaltenes are composed of highly poly-aromatic species with alkyl side chains

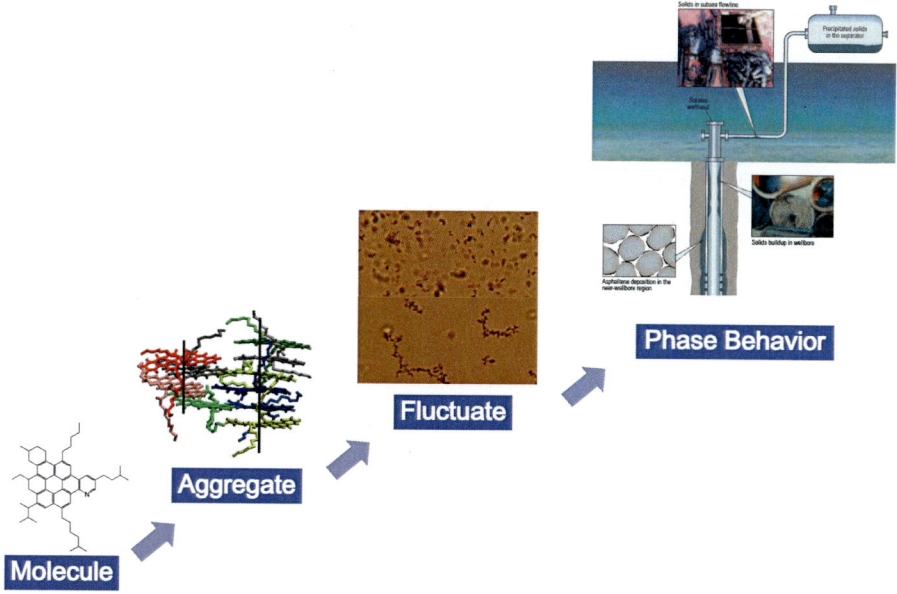

Fig. 1 Multilevel structure of asphaltenes from molecule to deposit phase. Asphaltene molecule has highly condensed aromatic rings and forms aggregate in both dilute solvent and crude oil reservoir. Flucates occur if the solvent environment cannot stabilize the aggregate, which then results in oil production problem such as well clogging. Asphaltene chemistry, including molecular structure, aggregate size, fluctuation kinetic, and phase equilibrium, has been widely concerned (modified from [7–10])

but also heteroatoms (N, S, and O) and metals (Ni and V), which make them the most polar and heaviest petroleum fraction. Because of the presence of heteroatoms in the aromatic structure of asphaltenes, these species are involved in a great variety of physical and chemical interactions, which lead to intermolecular association in asphaltenes and formation of aggregates. If the condition is thermodynamically instable (such as in poor solvent), the aggregation will continue and asphaltene fluctuation occurs.

The study and analysis on asphaltene is phenomenological. The complexity of asphaltene is manifested in two aspects: the numerous species involved in the system and the different form and size of molecular assembly. The complexity hindered the direct analysis on asphaltene, and the research on the asphaltene structure and aggregation is still an unfolding story. Recently, due to the progress in analytical technology and development of novel modeling and thermodynamic theory, the molecular structure of asphaltene composition and intrinsic nature of asphaltene aggregation is revealing. The present review is focusing on the structure and association behavior of asphaltene. The recent finding describing the process from asphaltene molecule to massive phase problem is summarized.

2 Chemical Structure of Asphaltene Molecule

2.1 *Molecular Weight (MW)*

Molecular weight is a very basic chemical property of a material. Although asphaltene is of considerable interest, its molecular weight remained controversial for a long time. Almost all the analytical measurement methods are affected by asphaltene aggregation. Since the formation of asphaltene nanoaggregates occurs at very low concentrations, the measured MW will be higher than the exact value for a monomer with the presence of aggregation. Distinguishing monomer and aggregate is challenging, and thus the debate on macromolecule or small-molecule size nature of asphaltene continued for decades.

Traditional VPO result is highly affected by asphaltene association. Vapor pressure osmometry is a very popular method and is capable of measuring the number average molecular weight (NMW) of chemicals and mixtures [11–13]. It converts the saturated pressure change of solution to NMW by calibrating with standard compounds. VPO result is structure independent since the colligative properties of dilute solutions only depend on the concentration but not the type of solute. VPO is mainly applied on the NMW determination of light (gasoline) and middle distillate (VGO) fractions. Many early works have extended VPO method to heavy petroleum, especially asphaltenes. This resulted in a very large NMW data which gives a strong support for the traditional macromolecule theory. However, asphaltene VPO result varied with type of solvent, temperature, and concentration [14]. At higher temperature, lower concentration, and solvent with higher polarity, the measured NMW is much lower. This phenomenon is related to asphaltene association as the aggregate formation varied with experimental conditions. Although many efforts have been made to lower the degree of aggregation of VPO, it is now generally agreed that under the restriction of dynamic range of VPO signal and polarity of solvent, elimination of the asphaltene aggregation to get NMW of true molecules is impossible.

Recent application of association model-based VPO data processing method indicates that the NMW of asphaltene monomer is small. Since VPO provides the apparent NMW of the combination of monomer and aggregates at different concentrations, the monomer NMW and aggregation kinetic/equilibrium parameters can be solved with a proper aggregation model. After the early report on empirical curve fitting, Yarranton group firstly proposed a linear polymerization asphaltene association model and used it in VPO data regression [15, 16]. The method was then further improved by increasing the component number [17, 18]. Recently, Zhang et al. proposed a hindered stepwise aggregation model for self-adaptive VPO data processing and measured NMW of light and heavy fractions of petroleum [19, 20]. The new method is applicable to mixture with both low and high degree of association, as shown in Fig. 2. Aggregation model-based VPO method reported a much lower NMW of asphaltene compared to traditional linear regression method. The NMW of asphaltene monomer by new method is in the range of

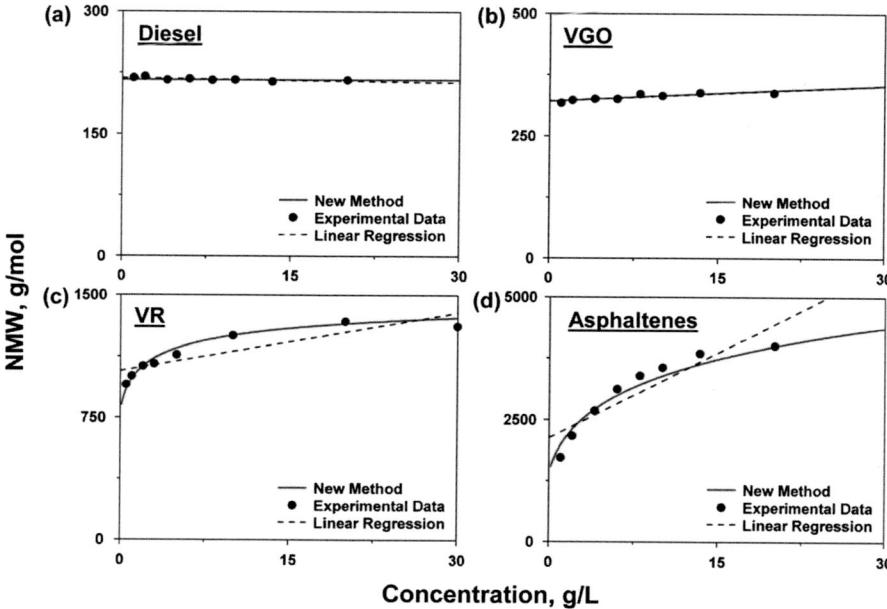

Fig. 2 Experiment data and prediction from the new analysis method and traditional linear regression for four petroleum fractions (**a**: diesel, **b**: VGO, **c**: VR, **d**: asphaltene). Linear regression only shows good agreement with the experiment for low-aggregate diesel and VGO fractions. For high-aggregate VR and asphaltene fractions, obvious derivations were observed (reprinted from [19])

~1,500 to ~1,800 Da. The exact number varied with association model applied and the origin of the asphaltenes. The recent progress on VPO measurement reveals that the large MW result by traditional VPO is incorrect. The asphaltene molecular size is much lower than previous estimation. However, it is notable that the new VPO method is not precise. Even after fractionation, asphaltene is an ultrahigh complex mixture of hydrocarbon and heteroatom-containing species. The single-component (Zhang et al.) or multicomponent (Yarranton et al.) association model cannot cover all the structural and interaction polydispersity. The value of association model-based VPO method is the calculation of association parameter, and thus the component interaction and aggregate size distribution can be quantified.

Mass spectrometry is another method that has been widely used to measure the molecular weight distribution. The early reports from LDI or MALDI is controversy as the asphaltene MW value varied by two orders of magnitude [21, 22]. The result from LDI and MALDI is different from traditional LDI or MALDI; two-step (L^2 MS) laser ionization eliminates the plasma-phase aggregation and desorbs asphaltene molecule at low-pulse energy. L^2MS is potential to analyze the MW distribution of asphaltene monomer. Pomerantz et al. for the first time performed two-step (L^2 MS) laser mass spectrometry analysis on a UG8 petroleum asphaltene [23]. They measured asphaltene MW distribution at various concentrations and experimental conditions. The result showed that the MW of asphaltene monomer

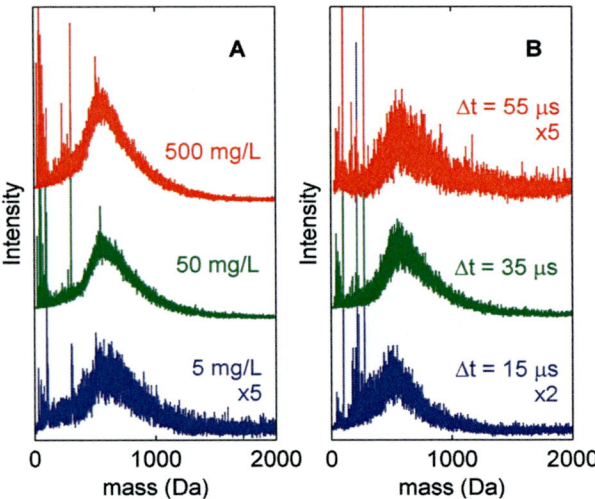

Fig. 3 L2MS mass spectra of asphaltenes at different (**a**) sample concentrations and (**b**) desorption–ionization time delays (reprinted from [23])

peaked at near 600 Da, and LDI mass spectral peaking at *m/z* higher than 1,000 should be the result of aggregation (Fig. 3). The petroleum asphaltene MW measured by L^2 MS is generally lower than 2,000 Da and averaged from 500 to 700 Da [24, 25]. The result from APPI and APCI FT-ICR MS shows similar MW range [26]. The MW distribution from FD MS is averaged at 1,238 Da and ranges from 300 to 2,500 Da, which is slightly higher than L^2 MS result [27]. ESI MS result shows a lower MW distribution [28].

The drawback of MS-based MW measurement is that it does not lead to quantitative information. The first reason is the ionization efficiency. As the ionization response for all species is not equal (e.g., ESI can only ionize polar species), the MW distribution by different MS methods are consequently not fully consistent. Moreover, the asphaltene aggregation cannot be fully eliminated under MS experimental conditions. McKenna et al. investigated the asphaltene aggregation in both parent crude oil and isolated fraction via APPI and ESI ionization sources and TOF mass spectrometer [29]. For toluene and methanol solvent system, asphaltene aggregates were observed even at very low concentrations (>50 μg/ mL). Moreover, they compared the MS-measured mass defect and expected (from bulk H/C analysis) and concluded that for all the atmospheric and laser-based (LD and L2) MS, a fraction of asphaltene molecule is absent since they form nanoaggregates. The result indicates that the MW distribution by MS-based method is not accurate but does not challenge the theory of small-molecule nature of asphaltene.

2.2 Detail Composition and Structure

Indicated by the elemental composition and ^{13}C or ^{1}H NMR analyses, asphaltenes are composed of highly fused aromatic rings with attached side chains. Heteroatom functional groups such as pyridinic and thiophenic unit were also included. The research was focusing on the size and number of fused aromatic ring systems. The early work by RICO observed a diacid product which indicated the presence of a bridged link between fused aromatic rings. The observation is in agreement with the traditional view that asphaltenes are large molecules with multi-core architectures [30, 31]. However, the RICO result still needs to be further investigated due to the complexity in the asphaltene composition and the reactions involved.

Optical spectra have suggested that the number of fused aromatic rings in asphaltene is ~7. After calculation of optical spectra of candidate model compounds by molecular orbital theory and comparing them with asphaltene data, the aromatic ring number in asphaltene molecules is presumed to be ranges from 3 to 15 with a peak at 7. The distribution was also used to calculate the full range (from near-infrared to ultraviolet) spectrum, and a good agreement between theoretical and experimental data was observed. The asphaltene size measured by high-resolution transmission electron microscopy (HRTEM) and scanning tunneling microscopy (STM) supports the result from optical spectra [32–34].

The progress in high-resolution mass spectrometry has facilitated the detail characterization of petroleum asphaltenes. High-magnetic field FT-ICR MS is capable of giving information elemental compositions of molecules, which are generally expressed in terms of DBE (ring plus double bond) and carbon number [35–37]. As asphaltene is the high polar fraction of petroleum, direct ESI FT-ICR MS analysis can be applied to characterize the polar species composition [28, 38, 39]. Mono-heteroatom-containing and multi-heteroatom-containing class species were observed. The polar species in asphaltene have high DBE value and low carbon-number distribution, which distribute near the structural boundary for fossil fuel aromatic compounds. After investigation by APPI FT-ICR MS, in which more types of species can be ionized, the asphaltene structural region was established. The boundary between asphaltene and maltene molecules is hydrogen-to-carbon ratio at ~1.1.

The fragmentation technology was applied to give information of asphaltene fused aromatic ring architectures. Collision-induced dissociations were performed at early report, but the results were not well understood due to the low resolution [40]. Very recent application of tandem FT-ICR MS result validates the presence of both single-core and multi-core motifs. Small DBE fragments were both observed from CID and IRMPD APPI FT-ICR MS spectra, which were produced bridge link dissociation (as can be seen in Fig. 4) [41, 42]. Hydrocarbon fragment ions were observed from CID ESI FT-ICR MS [43]. It was presumed to be the hydrocarbon cores that connected to heteroatom-containing cores. The results revealed that the heavy petroleums are not only composed of single-core molecules.

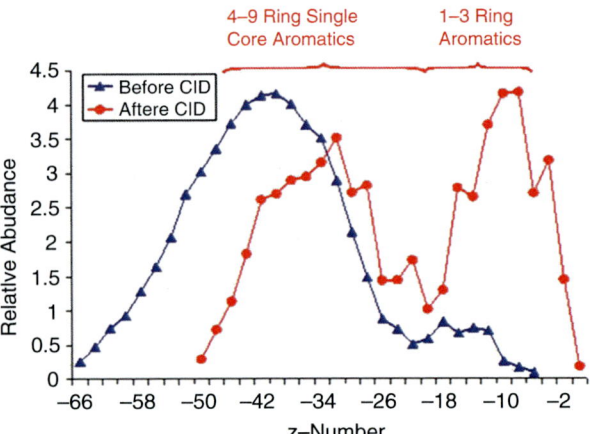

Fig. 4 Z-number distribution of the VR ARC4+ fraction before and after CID. Total concentrations in both cases are normalized to 100%. Y axis shows the relative concentrations of Z-number series (homologous series). Bimodal distribution indicates the presence of small aromatic pendants and large aromatic cores in the resid structure (reprinted from [41])

From recent investigation, it is clear that the asphaltene molecules have small molecular weights (<2000 Da) and are composed of highly fused aromatic rings and short side chains. Although there is still controversy, it can be presumed that asphaltene molecules are composed both of multi-core and single-core motifs. The latter one should be predominant according to the averaged structural analysis and the abundance of fragment ions.

3 Molecular Interactions and Aggregation Models

3.1 Asphaltene Molecular Interactions

Motivated by solving the problems encountered in heavy oil production, transportation, and refining, interactions between asphaltene molecules have been intensively studied during the past several decades. A variety of studies showed that the aggregates of asphaltenes and resins originated mainly from acid–base, hydrogen bonding, coordination, and π–π stacking interactions.

3.1.1 Acid–Base Interactions

According to the Brønsted acid–base theory, acids can interact directly with bases by proton transfer, acting as proton donors and acceptors. The acid–base interactions have been revealed in petroleum chemistry, especially for asphaltene molecules. Carboxylic acids can interact with aromatic nitrogen moieties by such action [44, 45]. The study of Gray et al. [46] implicated that ionic association represents the strongest interaction driving asphaltene aggregation [46]. Naphthenic acids are among the naturally occurring components in most crude oils. Asphaltene diffusion measured by NMR decreased with the addition of naphthenic acid, which indicates there are acid–base interactions between them [47].

3.1.2 Hydrogen Bonding

Hydrogen bond is a particularly strong dipole–dipole attraction when a hydrogen atom bound to a highly electronegative atom such as nitrogen and oxygen. For asphaltene molecules, groups such as hydroxyls and carboxylic acids can participate in hydrogen bonding with weak nitrogen bases, and nitrogen bases can associate with each other through mutual hydrogen-bonding interactions [48, 49]. Hydrogen-bonding interactions have been observed to play an important role in the precipitation of asphaltenes [50]. The simulation results revealed that alkylphenols tend to form H-bonds with –OH and –N groups of asphaltenes at their periphery sites. It is confirmed by high-resolution transmission electron microscopy (HRTEM) that octylphenol can reduce asphaltenes from precipitating to flocculate size [51]. The results of density functional theory (DFT) calculations show that hydrogen bonding is as important as π–π interactions for asphaltene aggregation [52].

3.1.3 Coordination

Oxovanadium and nickel compounds are widely distributed in asphaltenes, and these species contribute to interactions by axial coordination and hydrogen bonding. Axial coordination and hydrogen bonding are estimated as the preferred aggregation modes between metalloporphyrins and nitrogen-containing heterocycle fragments [53]. The relatively high correlation of the N/C ratio with vanadium and nickel contents in asphaltenes suggests that interactions of chelated porphyrin compounds are important in the asphaltene aggregation mechanism [50].

3.1.4 π–π Stacking

The face-to-face stacking, commonly called π–π stacking, arises from strong association of condensed aromatics in solution based on electrostatic and van der Waals forces. It is generally believed that the π–π interaction among poly-aromatic cores leads to their stacking, which is a critical driving force for asphaltene aggregation [7, 54–56]. The results of theoretical calculation indicate that the asphaltene–resin complex is very stable and dominated by π–π interactions [57]. Furthermore, dynamic simulations show that long side chains hinder the poly-aromatic core stacking [7, 55]. This interaction is rather controversial and ill defined [58], but it is commonly used to describe van der Waals and induction-type interactions between parallel aromatic ring systems [59].

However, Gray et al. [46] suggested π–π stacking is only one of many contributing factors driving the nanoaggregation. Other associative forces such as acid–

base interaction, hydrogen bonding, metal coordination, and hydrophobic pockets are likely to act simultaneously with π–π forces to drive asphaltene aggregation. It was also appreciated by Hoepfner and Klerk [60] that π–π ring stacking was just one of the possible aggregation forces present in asphaltenes. Study using UV–vis and fluorescence spectroscopies shows that the association of the majority of the vanadium and nickel petroporphyrins in crude oils with the asphaltene fraction may be due to other functionalities appended to the porphyrin ring, rather than favorable π–π stacking interactions of aromatic rings with the porphyrin core itself [61].

Asphaltene is a complex mixture comprising many components which indeed may vary from one crude oil to the other. Up to date, the exact composition and molecular structure of asphaltenes remain unresolved. As a result, the nature of interactions involved in asphaltene aggregation also remains controversial and certain asphaltenes might possess unique properties.

Generally, the single PAH core in asphaltene molecules is assumed to be the primary site of intermolecular attraction because of both its polarizability and some degree of charge separation associated with heteroatoms in the aromatic ring system. However, extensive studies indicate only a fraction of the interactions among asphaltenic molecules may take place through stacking of aromatic sheets [60–62]. Despite the large size of the aromatic group and the lack of steric interference from the peripheral side chains, the association of hexabenzocoronenes (13-ring condensed aromatics) was much weaker than the observed association of asphaltene [46].

Besides intermolecular interactions among asphaltene molecules, their aggregation behavior also strongly depends on the surrounding solvent. The forces between asphaltenes are thought to be dominated by van der Waals interaction in paraffinic solvents [63]. Using atomic force microscopy (AFM), Wang et al. showed that the forces between asphaltenes are highly sensitive to the composition of the solvent and the transition of force profiles from repulsive to a weak attraction as the solvent power gradually increased [64]. Porte et al. [65] proposed that aggregation and precipitation are presumably controlled by different intermolecular forces. Aggregation is induced by strong specific interactions (i.e., hydrogen bonds) in good apolar solvents, and precipitation is determined by weak nonspecific dispersion attractions in bad apolar solvents.

3.2 Structural Model of Asphaltene Aggregation

As stated above, asphaltene molecules can self-associate to aggregation particles due to several aforenamed interactions. Several association models which account for the experimental data have been proposed to interpret the different aggregation states and mechanism of formation.

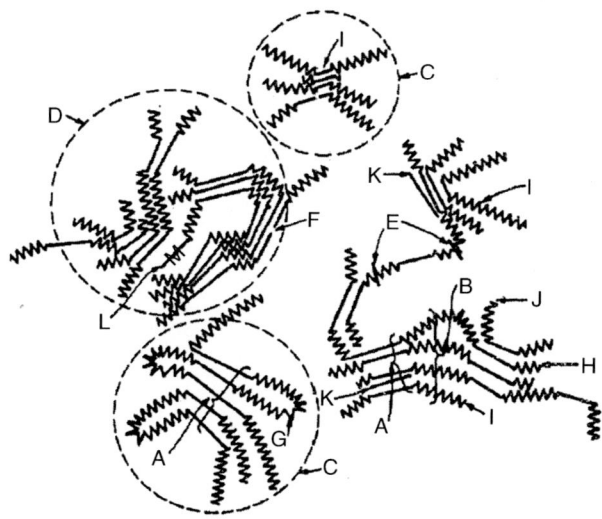

A. Crystallite B. Chain bundle C. Particle D. Micelle E. Weak link F. Gap and hole
G. Intracluster H. Intercluster I. Resin J. Single layer K. Petroporphyrin L. Metal

Fig. 5 Schematic representation of Yen model (reprinted from [54])

3.2.1 Yen Model

Mainly based on solid asphaltene characterization data from XRD measurements, Professor Teh Fu Yen and his coworkers [54] previously proposed a hierarchy of asphaltene structures to account for many of their characteristics known at that time. This hierarchical structure of asphaltenes has been termed the Yen model as shown in Fig. 5. It shows the aggregation mechanism of asphaltenes from the molecule state to the cluster state: individual aromatic sheet can be stacked to form elementary particles, and these particles can further associate micelles; asphaltene micelles can cluster into aggregates when the concentration is sufficiently high. The molecular association of stacked aromatic sheets in solution was driven by π–π interaction and hydrogen bonds. Contaminates, such as metal, can be an aid to micelle association. In spite of resins being similar to the corresponding asphaltene fractions in size, they have the lower degree of condensation. Therefore, resins and other aromatic compounds present in the oil can only associate with asphaltenes. The Yen model has been useful for bulk measurements on phase-separated asphaltenes.

3.2.2 Yen–Mullins Model

Stimulated by advances in analytical chemistry, asphaltene science has progressed dramatically in the last decade. Based on precepts of the Yen model, a new model

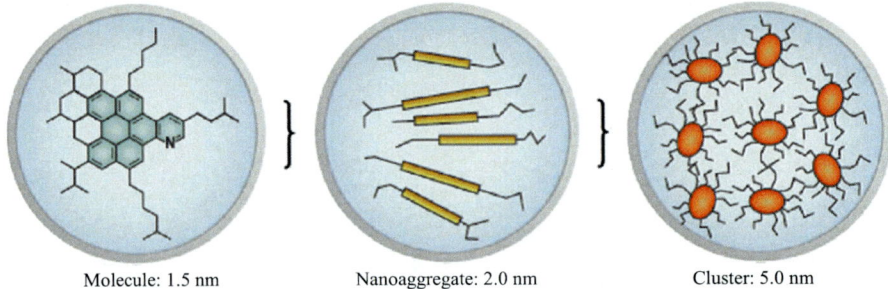

Molecule: 1.5 nm   Nanoaggregate: 2.0 nm   Cluster: 5.0 nm

Fig. 6 Schematic representation of Yen–Mullins model (reprinted from [66, 67])

has been proposed by Mullins to conform to enormous studies. This model shown in Fig. 6 has been called the modified Yen model [66, 67] and equivalently the "Yen–Mullins model" [40, 68, 69]. The basic features of the model are presented as follows:

1. Asphaltenes are dominated by the island molecular architecture with a single polycyclic aromatic hydrocarbon (PAH) ring system in the molecular core. They have the most probable molecular weight of 750 Da with a width of 500 to 1000 Da. The distribution of asphaltene PAHs is centered roughly at 7 fused rings.
2. Asphaltenes exist in crude oil as three distinct hierarchical structures: molecules, nanoaggregates, and clusters.

In this model, the predominant asphaltene molecular architecture consists of a single moderate-sized polycyclic aromatic hydrocarbon (PAH) ring system with peripheral alkane substituents. The PAH is the primary site of intermolecular attraction.

The asphaltene molecules can form nanoaggregates with a single, disordered stack of PAHs. The dimensions of these nanoaggregates are ~2 nm with aggregation numbers ~6. The exterior of the nanoaggregate is dominated by the alkane substituents. They do not need resins to form or to be stably suspended. The heaviest resins participate in the aggregation at a level of ~15% mass fraction.

The nanoaggregates can further form clusters of nanoaggregates in crude oils and solvents. These asphaltene nanoaggregate clusters are not much bigger than the nanoaggregates, and aggregation numbers are estimated roughly to be eight nanoaggregates. The clustering can be strongly impacted by changing temperature, concentration, and liquid-phase properties.

3.2.3 Gray's Model

Both the Yen model and Yen–Mullins model stressed the formation of clusters of molecules by π–π stacking of aromatic rings. Recently, Gray et al. [46] demonstrated that π–π stacking is only one of the interactions and proposed a new

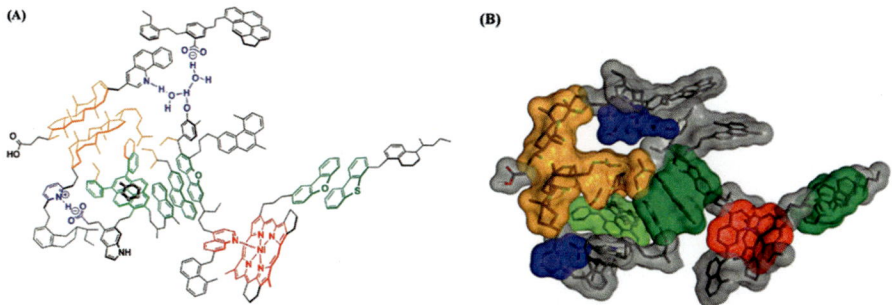

Fig. 7 Schematic representation of Gray's model (reprinted from [46]). Associations between molecules are color-coded in (**a**) the molecular depiction and (**b**) the space-filling version: acid–base interactions and hydrogen bonding (*blue*), metal coordination complex (*red*), a hydrophobic pocket (*orange*), and π–π stacking (face to face, *dark green*; within a clathrate containing toluene, *light green*)

supramolecular assembly model (cf. Fig. 7) for aggregation of asphaltenes. The driving forces for this model comprise acid–base interaction, hydrogen bonding, metal coordination, hydrophobic pockets, and π–π stacking. The implication of this supramolecular assembly of complex molecules is that the dispersed nanoaggregates of asphaltenes in solution are porous structures

3.2.4 Yarranton's Model

The asphaltene VPO molecular weight has been shown to increase with asphaltene concentration until a limiting value is reached [15]. The limiting molar mass depended on the temperature and solvent. Base on the above experimental work, Yarranton and his coworker [16] proposed a stepwise polymerization model of asphaltene association, in which asphaltene self-association was modeled in a manner analogous to linear polymerization (cf. Fig. 8). The key concept in the model is that asphaltene molecules may contain single or multiple active sites (functional groups) capable of linking with other asphaltenes. Asphaltenes consist mainly of propagators but also contain a small proportion of terminators. Resins consist mainly of terminators but also contain a small proportion of propagators. Molecules with multiple active sites act as propagators, and molecules with single active sites act as terminators in process of association. The type and strength of the potential links may vary considerably because asphaltenes consist thousands of chemical species featuring a variety of functional groups. The results of the model calculation were validated with experimental data from their further studies [16, 70].

Fig. 8 Schematic representation of Yarranton's model (reprinted from [16])

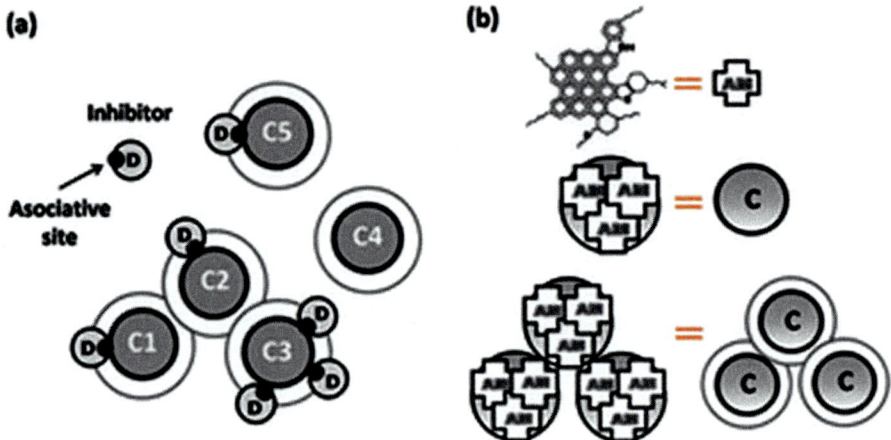

Fig. 9 Schematic representation of PAC model (reprinted from [71]). (**a**) Two-dimensional sketch of the bonding topology of the PAC model. (**b**) Terminology used by the model, where AM is an asphaltene molecule and C is a colloidal particle

3.2.5 PAC Model

Barcenas et al. [71] took diverse structure of asphaltenes into account and proposed a new simple model of particle aggregation control (PAC) to study the influence of associative inhibitors in the colloidal clustering (cf. Fig. 9). The PAC model is sufficiently flexible to describe mean cluster size and cluster size distribution of the asphaltene colloid aggregates of different origins [72, 73]. Similarly to experimental results of Agrawala and Yarranton [16], theoretical calculation by the PAC model revealed that the increase of resin concentration leads to reducing of asphaltene cluster size.

In summary, the presence of asphaltenes in crude oils is of considerable relevance for a range of stages from oil recovery via transport to refining. The knowledge of the interactions between asphaltene molecules/aggregates in an organic solvent throws light on the ultimate understanding of asphaltene aggregation behavior in crude oil and in solvents. The lack of a precise molecular identification of each individual asphaltene component hinders the full understanding of the interactions between asphaltenes, their aggregation, and the subsequent precipitation of these substances.

4 Asphaltene Deposition

Asphaltenes are the heaviest and most polar fraction of crude oil. One important property of asphaltenes is their tendency to self-associate and form aggregates of colloidal dimensions both in the native crude oils and in solution. Despite vast work devoted to understanding asphaltene molecular structure and the mechanisms of asphaltene precipitation and deposition, the latter still represents a challenging problem.

At early stages, Nellensteyn worked on the properties of asphalts and proposed the conception of asphalt micelle [74]. This colloid conception has been further developed by Pfeiffer and his collaborators [75], who suggested that asphaltene molecules are the centers of micelles and dispersed by resin molecules, which are a part of maltenes. The other part of the maltenes is oily constituents, which surrounded the micelles. Yen and his coworkers [54] proposed a hierarchy of asphaltene structures termed the Yen model. It shows the aggregation mechanism of asphaltenes from the molecule state to particles, micelles, and at last the cluster state.

The principal mechanism of asphaltene deposition seems to be self-association and then desolubilization due to changes in the environment to which they are subjected. As conditions change, the molecules undergo aggregation, coagulation, or flocculation to form macroscopic particles, which then precipitate from solution. It reveals a hierarchy of asphaltene aggregations as conditions of environment occur to change.

4.1 Asphaltene Micellization and Aggregation

Because of the amphiphilic nature of asphaltenes, aggregation of asphaltene molecules was regarded as formation of micelles a decade ago. Critical micelle concentration (CMC) or critical aggregation concentration (CAC) was proposed to define the value at which asphaltenes aggregation starts. There had been several reports claiming that the CMC can be determined by using a variety of different methods such as surface/interfacial tension, calorimetric titration, or spectroscopy.

Sheu and coworkers [76] measured the surface tensions of asphaltenes extracted from Ratawi vacuum residue. The surface tensions were plotted as a function of asphaltene concentrations (cf. Fig. 10). They defined the discontinuous point as a CMC. Similar discontinuities or breakpoints in interfacial tension or calorimetric titration measurements have also been observed and regarded as CMC values.

The CMC value of asphaltenes in toluene determined by these methods is around several g/L [77–79]. Surface and interfacial tension data reported by Mohamed et al. indicate that CMCs measured for the n-pentane insolubles (C5I fraction) in toluene and pyridine solutions were consistently higher than those in systems containing the n-heptane insolubles (C7I fraction), indicating a lower association

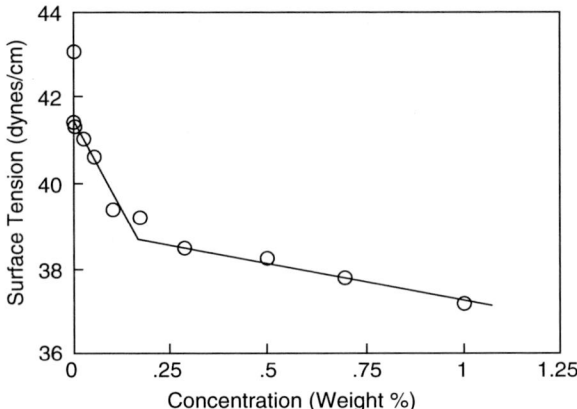

Fig. 10 Surface tension as a function of concentration for asphaltene in pyridine at $T = 25°C$ (reprinted from [76])

tendency in the C5I fraction [80]. Calorimetric measurements indicate that asphaltenes follow a stepwise mechanism followed by a steady level above the apparent CMC [77]. The CMCs fell in a narrow range between 3.2 and 4.9 g/L toluene for several n-heptane asphaltenes.

Oh et al. [81] used near-infrared (NIR) spectroscopy to determine the onset of asphaltene precipitation by heptane titration. In their study, distinct breakpoints in the plot of NIR precipitation onset versus concentration were observed and interpreted to be CAC values. CACs for asphaltenes dissolved in toluene, trichloroethylene, tetrahydrofuran, and pyridine observed at concentrations of 3.0, 3.7, 5.0, and 8.2 g/L, respectively. CMC values of asphaltenes determined from surface tension measurements were slightly higher than the CAC values measured by NIR onset measurements.

However, the interfacial tension experiments on Athabasca asphaltenes indicated that no micelles are formed [15]. Merino-Garcia and Andersen [82] also did not observe a CMC or CAC of asphaltene in organic solutions. Meanwhile, Andersen and Speight [83] reported that resins and aromatics do not show self-aggregation behavior. The properties of the asphaltene [50] and different extraction procedures [70] have shown influence on the association behavior.

4.2 Asphaltene Nanoaggregation

More recently, a large amount of data reveals that asphaltenes can self-associate and form colloidal aggregates in toluene solution at concentrations lower than CMC. The particles are most commonly described as "nanoaggregrates" with their dimensions of 2–10 nm. As stated by Mullins [66], after several asphaltene molecule aggregates, the nanoaggregate projects primarily the steric-repelling alkanes to the outside world, and additional asphaltene molecules are not able to achieve close approach to the interior PAHs. Thus, additional asphaltene molecules form new nanoaggregates of small aggregation numbers and not large nanoaggregates.

A large number of techniques have been used to investigate asphaltene aggregation and have provided consistent results. High-Q ultrasonic studies were the first to correctly determine the CNAC of asphaltenes in toluene and showed asphaltene CNACs at ~100 mg/L [84]. Direct-current (DC) electrical conductivity [85], NMR [86], centrifugation [87, 88], and fluorescence methods [89] are all exhibiting similar CNAC values. Furthermore, the results of high-Q ultrasonics and direct-current (DC) electrical conductivity indicate that different asphaltenes have somewhat different values of CNAC [84, 85]. Both techniques show the CNAC for UG8 asphaltene is around 150 mg/L, while the CNAC for BG5 asphaltene is 60 mg/L.

To be noted, the critical nanoaggregate concentration (CNAC) does not define the onset of aggregation but rather defines the concentration where further growth of the nanoaggregate shuts off. Therefore, CNAC concentration is larger than initiation aggregation concentration, which is the onset of aggregation.

4.3 Clustering of Asphaltene Nanoaggregates

The asphaltene nanoaggregates can further cluster in crude oils and toluene. At higher concentrations than the CNAC, a secondary aggregation process known as the clustering of nanoaggregates takes place [67]. These clusters can form with a length scale of several nanometers and larger and have small aggregation numbers [87].

The critical clustering concentration (CCC) is observed at higher concentrations by using both DC conductivity and centrifugation methods. The formation of clusters of asphaltene nanoaggregates has been detected at concentrations of roughly 2 g/L in toluene by DC conductivity [22, 23]. As stated by Mullins, the CMC is likely responding to the clustering concentration [8]. The CCC values have also been measured by the kinetics of floc formation on normal alkane addition to asphaltene–toluene solutions.

The nanoaggregates can undergo coagulation or flocculation to form macroscopic particles when the colloidal environment changes. Recent studies indicate that the aggregates are porous and can trap solvents [90, 91] and low MW compounds [92, 93]. Diverse experimental methods have shown that asphaltene aggregates form loose porous structures such that the solvent molecules can diffuse through them.

The study of Derakhshesh et al. [91] indicated that PAHs pyrene and phenanthrene were significantly occluded within asphaltene nanoaggregates. Roughly 30–50% (v/v) of solvent entrainment has also been found from the calculation of average aggregate molar masses [94]. The H_2O_2/CH_3COOH system was applied to release the occluded compounds from asphaltene aggregates and a series of unsaturated alkenes and terpene (hopenes) compounds detected inside the aggregates [95, 96]. It indicates that the macromolecular structures of asphaltenes have protected these compounds from being influenced by the alteration processes in the oil reservoirs and enable them to survive over geological time.

Coupling scattering experiments and viscosity measurement, a fractal model can be used to describe asphaltene aggregate structure in strong solvents, which in consequence were trapped partly [90]. NMR results show relaxation enhancement of the ambient oil because asphaltene is directly proportional to the relaxation rate of the given oil component. A porous asphaltene model was proposed and is consistent with the experimental results [97].

In order to explore the selectivity of the entraining solvents, a mixture of good and poor solvents has been used in entrainment experiments [98]. The results show the preferential enhancement of the toluene entrainment within aggregates in both the Decalin/toluene and *n*-heptane/toluene solvent pairs. Furthermore, the total solvent entrainment increases with increasing aggregate size of asphaltenes.

Despite of the many studies supporting loose structure of asphaltene aggregate, compact aggregates with the order of size of 1 μm have been found by SAXS. Further study using scanning electron micrographs confirmed the presence of dense spherical structures of the asphaltenes. Based on the above results, Sawidis et al. [99] make a conclusion that the structure of flocculated asphaltenes is clearly a compact organization of entities ranging from nanometers to microns.

5 Size Distribution of Asphaltene Monomers and Aggregates

5.1 Size of Asphaltene Monomers

Asphaltene is the heaviest fraction of crude oil composed of a broad distribution of species that are polydisperse at the molecular level. The heavy crudes are characteristically more difficult to process than petroleum distillates because of larger molecule size and higher amounts of heteroatoms. Stimulated by the need for production and processing heavy petroleum feedstocks, there has been considerable interest in the chemical composition and structure of asphaltenes. Despite great progress in chemical identification of asphaltenes, there is still some debate about the size of these feedstocks due to their complicated compositions and tendency to aggregation.

There are three primary methods used to obtain asphaltene dimension, direct molecular imaging, interfacial property, and molecular diffusion. Scanning tunneling microscopy has been applied to investigation of asphaltenes structure at early stages. It reveals that the bulk of the PAHs have a long axis in the range around ∼1 nm, which would correspond to roughly seven fused rings [100]. Direct molecular imaging by high-resolution transmission electron microscopy shows that the bulk of the polycyclic aromatic hydrocarbon ring systems has a length of scale slightly larger than 1.0 nm [33]. However, the images representing the aromatic ring systems of asphaltenes due to the aromatic portion are readily imaged, while the alkanes are not readily observed. Furthermore, it is difficult to distinguish aggregates versus individual structural units because of the heterogeneous nature of asphaltenes.

The average diameter for asphaltenes adsorbed at the interface has also been estimated from the areas [78, 80]. The estimated average radii for asphaltenes at the interface are also in agreement with the values mentioned above. However, the concentrations used in interfacial property measurements are larger than 1 g/L, which are above a given threshold value for nanoaggregation. Then, the asphaltene aggregation might occur as stacking on the interface.

Fluorescence depolarization (FD) [101, 102], fluorescence correlation spectroscopy (FCS) [103], and nuclear magnetic resonance (NMR) [86, 104] techniques were performed on molecule diffusion measurements for asphaltenes, and the hydrodynamic dimension of diffusing species is evaluated. Basic agreement among these methods indicates that the average diameters of petroleum-derived asphaltenes range from 1.0 to 3.0 nm. The translational diffusion of the asphaltene has also been determined by using the Taylor dispersion technique. The sphere-equivalent hydrodynamic diameter of the coal and petroleum asphaltenes was estimated to be 1.1–1.6 nm [105–107]. The combined intrinsic viscosity and diffusion data suggest that the shape of coal asphaltenes can be modeled as prolate ellipsoid.

The study by membrane diffusion [108] also provided consistent dimension of asphaltenes. Recently, membrane diffusion experiments have been performed on residue and its fractions [109]. Obvious polydispersity for each fraction has been presented by their size distribution. The average hydrodynamic diameter of the end cut was estimated to be 4.7 nm, as opposed to a range of 1.1–1.7 nm for the four narrow fractions in size. Strong tendency of asphaltenes to aggregate suggested that the large size of the end cut results from the aggregation of asphaltene molecules. Further study showed that the hydrodynamic diameter of sulfur-containing monomers of SFEF fractions ranges from 0.74 to 1.45 nm at concentration of 1 g/L, as shown in Fig. 5.1 [110]. The size of maltene monomers spans a range of 1.87 nm to 2.29 nm at 0.1 g/L and presents a more significant polydispersity than SFEF fractions. The size variation of the SFEF fractions and maltenes to yields demonstrates a continuous distribution in size for petroleum residue.

5.2 Size of Asphaltene Aggregates

The recent advances in asphaltene science have led to a consensus that there is a strong tendency for asphaltenes to self-association. Aggregation of asphaltene molecules has been studied extensively by NMR diffusion [86], direct-current (DC) electrical conductivity [85], and high-Q ultrasonic spectroscopy [84], centrifugation [88], and mass spectroscopy [111]. All experimental results to date

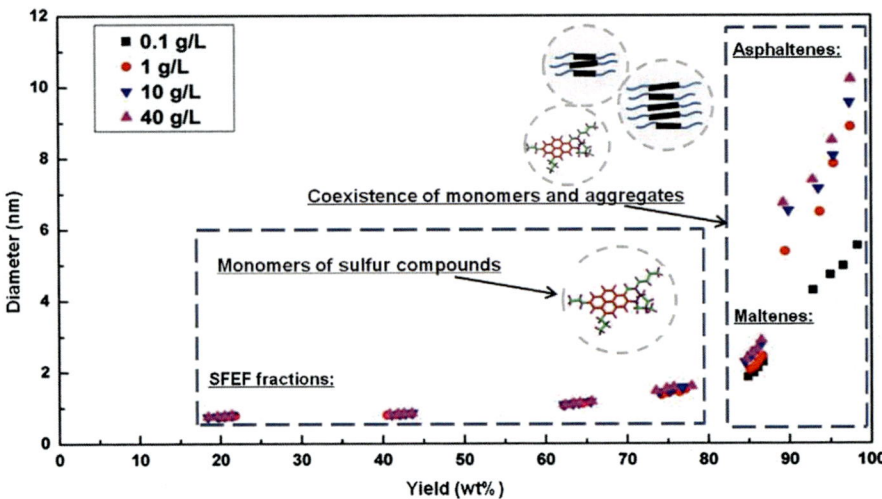

Fig. 11 Size distribution of monomers and aggregates of sulfur-containing compounds in VAR fractions (reprinted from [110])

confirmed that asphaltenes in toluene form nanoaggregates at low concentration of roughly 150 mg/L.

The average diameters of the asphaltene nanoaggregates obtained by the forenamed methods mostly ranged from 3 to 10 nm [86, 112–118]. Complementary SAXS and SANS results give a total radius of 3.2 nm and an average core radius of 1.8 nm for the nanoaggregate [114]. The mean diameter of the nanoaggregates for different asphaltenes assessed by Rayleigh scattering is in a range of 5–8 nm for concentrations from 30 mg/L to 10 g/L at ambient conditions [118]. These data are in very good agreement with the 5–9 nm size range obtained using membrane diffusion measurements [119]. These results were also confirmed by atomic force microscopy [116] and direct-current (DC) conductivity [117].

Optical absorption and fluorescence emission analysis coupled with molecular orbital (MO) calculations indicated that the most probable number of fused rings is seven [120, 121]. A comparison between the L2MS and SALDI–MS results reveals that yields the most probable aggregation number observed for nanoaggregates of petroleum asphaltenes is approximately 6–8 molecules [111]. Incorporation of Raman spectroscopy measurement and the X-ray data gives rise to an estimate of eight asphaltene molecules in each asphaltene aggregate [122]. Using vapor pressure osmometry (VPO), Yarranton et al. [15] obtained lower asphaltene aggregate number of 2–6. However, a wide range of aggregate number of 1–30 molecules per nanoaggregate has been represented by Yarranton' group [123].

More recently, the dimensions of sulfur-containing compounds in residua were obtained from membrane diffusion measurements (cf. Fig. 11) [110]. Asphaltene aggregates cover the span of diameters from about 4.29 to 5.54 nm at concentration of 0.1 g/L. In spite of coexistence of molecules and aggregates, maltenes are dominated by monomers and range from 1.87 to 2.89 nm at concentrations larger than 0.1 g/L. However, aggregates dominated in asphaltenes with the size distribution of 4.29–10.20 nm over concentrations of 0.05–40 g/L. The average diameters of asphaltene fractions decreased to 4.02 and 3.34 nm at concentrations of 0.05 and 0.03 g/L, respectively. Therefore, it reveals that aggregation of asphaltene molecules can occur at concentration lower than 0.1 g/L.

NMR diffusion measurements of asphaltenes were carried out over a wide range of concentrations [124, 125]. The diffusivities remain constant at a dilute state. As the concentrations increase, there is a transition of aggregation and two different aggregates were observed. Kawashima et al. [104] have also obtained different aggregates with multi-scales. At low concentrations (0.1 and 1 g/L), small (diameter of 1.2–1.4 nm) to medium aggregates (7.0–8.4 nm) were observed, and at higher concentrations (10 and 30 g/L), only medium aggregates (4.8–7.6 nm) were detected. Headen et al. [126] also highlighted by SANS the coexistence of big aggregates with smaller ones.

Yarranton et al. [127] applied a variety of experimental techniques to a single-source asphaltene sample at the same conditions to investigate the size of asphaltene monomers and aggregates. The membrane diffusion, Rayleigh scattering, and diffusion experiments give a similar diameter which range from 5 to 9 nm. However, the size determined from nanofiltration experiments and SAXS data is larger than the diameter mentioned above.

Furthermore, experimental results have presented that an aggregate can form clusters as environments change. It induced appearance of cluster particles, which have larger size than the aggregates [128, 129]. Small-angle X-ray scattering (SAXS) and small-angle neutron scattering (SANS) have been extensively performed to investigate the colloidal structures of asphaltene [112–115, 126]. Despite of several studies that provide consistent results with other methods of 3–10 nm for asphaltene nanoaggregates [112–115], much larger dimensions have been presented [126, 130]. The gyration diameter of asphaltene aggregates has even been detected to be approximately 0.45 μm [126]. Hydrodynamic radii of clusters of light asphaltene fractions are between 5 to 10 nm, whereas a small fraction of asphaltenes is made of larger clusters with radius of around 40 nm [111].

As stated by Mullins [66], SAXS and SANS measurements are sensitive to the radius of gyration, while diffusion and transport terms are sensitive to the hydrodynamic radius. Measurement of their sizes from gravitational gradients is sensitive to the physical radius. Furthermore, various model fits were applied to the scattering curve in order to study the morphology of the nanoscale aggregates [131]. It reveals that the gyration radius differs significantly among these models, and the polydisperse radius oblate cylinder model best approximates the shape of asphaltenic aggregates. As a result, the different dimensions impede direct comparison of nanoaggregate size determined by different methods.

5.3 Effects on Size of Asphaltenes

As stated above, the changes of environmental conditions can cause different aggregation states to occur. Many factors, such as temperature, types of solvent, and additives, can contribute to the changes of aggregation state.

The effect of temperature on asphaltene aggregation has been performed by several methods. The results of SANS and SAXS measurements show that there is a decrease in nanoaggregate size with increasing temperature [132–134]. At high temperatures, reversible aggregation of asphaltene leads to stable small entities. When decreasing the temperature, irreversible aggregation of asphaltene occurs, corresponding to a large increase of the aggregate size. It is validated by the recent experimental results of Eyssautier and coworkers where the radius of gyration of asphaltene aggregates decreases from 8.2 to 6.5 nm from 80 to 240°C [135]. However, the results of Headen et al. [126] and Hurtado et al. [115] show that there is no monotonous change for the gyration diameter of asphaltene aggregates with temperature. Temperature-dependent study of asphaltene dimension shows fractions A1 (lower solubility) have a minimum size during the whole heating stages [136]. When temperature is raised, a fraction of molecules at the asphaltene colloidal periphery dissolves, thus reducing size of asphaltenes; however, a further increase in temperature thins the solvent layer, promoting flocculation and thus increasing the size of them.

Asphaltenes are defined by their soluble behavior in the presence of solvents. Therefore, the solvents have significant influence on their solution characterization and phase behavior. Asphaltenes have been precipitated into two fractions, and SANS showed that the second fraction (addition of n-pentane at a high solvent-to-oil ratio) formed aggregates with lower radius of gyration [137]. Using the same method, asphaltene aggregates have been found to appear as stable solvated entities with radii of gyration of a few nanometers at asphaltene volume fractions ranging from 0.3 to 3–4% [133]. The results from Hurtado et al. [115] indicate that aggregate radii are unaffected by dilution with 1-methylnaphthalene and increase significantly on dilution with n-dodecane. When the heptane fraction in solvent is increased, two times increase in gyration radius has also been found by using viscosimetric and SANS experiments. However, adsorption isotherm measurements [138] show that the native resins exhibit a high asphaltene dissolution power and decrease the size of asphaltene aggregates. Furthermore, asphaltenes have been found to expand by approximately a factor of 4 when in contact with liquid toluene [127]. It needs to be careful when comparing the size of asphaltene and its aggregate from the results of different experiments.

One method for controlling the precipitation of asphaltenes is the addition of dispersants capable of stabilizing the colloidal suspension of asphaltene nanoaggregates. Such chemical additives generally can prevent the formation of precipitate [72] and lead to a larger proportion of asphaltene nanoaggregates of smaller sizes [139].

Furthermore, ultrasound and heat treatments were employed in an attempt to completely disaggregate the asphaltene nanoaggregates in solution at room temperature. The results show that asphaltenes could not be disaggregated completely in the study of Derakhshesh et al. [118]

In summary, while it is well established that asphaltenes self-associate, the size distribution of the asphaltene nanoaggregates remains uncertain due to the vast complexity of constituents. Diverse experimental methods indicate that the size of petroleum asphaltene monomers ranges from 1 to 3 nm. Due to structural constraints, the asphaltene molecules may form nanoaggregates that apparently have a limited number in crude oils and solution. The size of the nanoaggregates has a range of 3–10 nm. Direct comparison of asphaltene monomer and aggregate size from different studies is difficult because of the complex composition of asphaltenes. As a result, the aggregation study should take into it consideration.

6 Molecular Simulation

6.1 Atomic Level

Molecular dynamic (MD), molecular mechanical (MM), and quantum mechanical (QM) simulation have the capacity of modeling asphaltene aggregation systems at the atomic level. The chemical detail of the simulation methods is not the same so as to the accuracy of the result. Since the computation power is limited, all the atomic-level simulations can only be conducted on a small- to medium-size scale and on very short time scale. Although the result cannot cover the chemical complexity of the asphaltene system, the molecular simulation is still very useful as it gives some information on the detailed structure and association energy of the aggregate system. The simulation can be seen as a "computer experiment" which predicts the result that cannot be provided by traditional laboratory experiments. It will help to understand the nature of asphaltene aggregates and give guidance to eliminate or predict the asphaltene deposition.

Due to the limitation of analytical measurements, the aggregate configuration is very hard to obtain from experimental data. One of the most useful aspects of molecular simulation is the information on the detail structure of asphaltene aggregates. From molecular simulation under vacuum, aromatic ring stacking occurred for both large- (22 aromatic rings) and small-island (Yen–Mullins model) asphaltene model compounds [140, 141]. The result shows that asphaltene model compounds form three types of aggregates, which are face to face, offset stacking, and T shaped. A snapshot structure is shown in Fig. 12. By changing the structure of input asphaltene molecules, the calculations showed the steric hindrance effect of the side chains on the aggregates [140]. The calculations of Jian et al. revealed that long side chains have a negative effect on the aromatic core stacking and also favor aggregation through hydrophobic association

Fig. 12 Snapshot of MD simulation of six molecules of asphaltene model compound in toluene at 300 K at point I (at ~3 ns) showing the formation of two trimers. Toluene molecules have been removed for clarity. Parallel and nonparallel stacking of aromatic cores is observed (reprinted from [142])

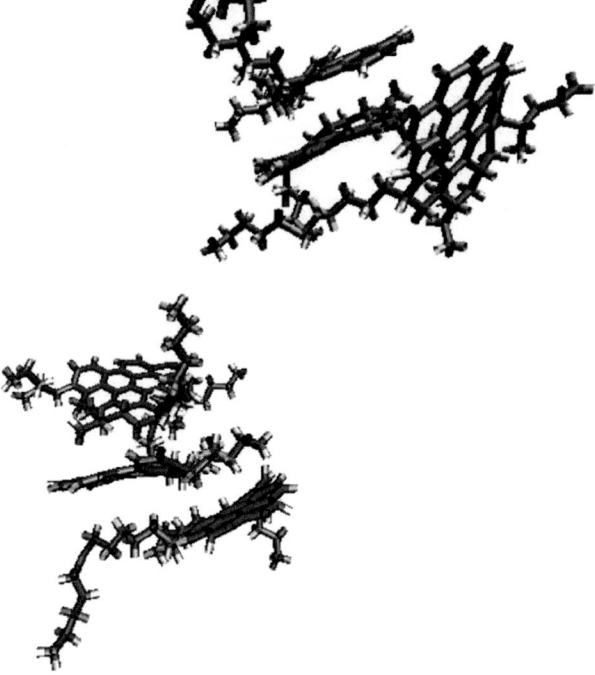

[7, 143]. Moreover, the effect of temperature and solvent effect can also be simulated. The published simulation result showed that with increasing temperature, the distance in the asphaltene dimer decreased [144, 145]. The simulation on asphaltene molecules with different types of solvent showed that the aromatic core stacking distance in a bad solvent (heptane) is shorter than in a good solvent (toluene) [145]. The asphaltene interaction in supercritical CO_2 was also simulated [146].

Another useful result from molecular simulation is thermodynamic data for the asphaltene aggregate. The density functional theory (DFT) calculation has revealed that in asphaltene model compounds the binding energy is about 60 KJ/mol [147]. Another report by coarse-grained model that with interaction potential derived from molecular mechanics has shown that the binding energy is ~130 KJ/mol [148]. The binding energy from a different model is not the same since the simulation methods and the model asphaltene structure are different. A recent study by Sedghi et al. revealed that the association energy increases substantially with the number of aromatic rings [56]. They also concluded that the heteroatoms attached to aromatic cores, other than those attached to side chains, have a large effect on the association free energy. A more detailed DFT calculation was conducted later by da Costa et al., and the association free energy of different types of aromatic hydrocarbons as asphaltene model compounds has been evaluated [149]. The contribution of hydrogen bonding and aromatic ring stacking is investigated by da Costa et al. from DFT calculations [52]. By comparing the ΔH and ΔG^{298} values, they concluded that the hydrogen bonding is as important as aromatic ring stacking during asphaltene aggregate formation. There were also predictions of other properties by molecular simulation, such as density and solubility [150–152].

Besides the investigation on the asphaltene aggregation process, the molecular simulation method has also been applied to understanding the interaction between asphaltene and other species, such as resins, water, and inhibitors. Alvarez and Ramirez evaluate the interaction potential curve between asphaltene–asphaltene, asphaltene–resin, and resin–resin systems by DFT [147]. Ortega-Rodríguez et al. used QM/MM to simulate the medium effect of asphaltene–resin mixture and summarized the aggregation tendency against the polarity of the solvent (heptane, toluene, and pyridine) [148]. Mikami et al. used heptane and toluene as representations of oil and simulate the microstructure of asphaltene at the oil–water interface [153]. The results reveal the different asphaltene behavior at low and high abundance and clarified the formation of an asphaltene film at the oil–water interface. The charged and uncharged asphaltene behaviors were investigated, and anion carboxyl asphaltene formed both stacked and T-shaped aggregate at oil–water interface [154, 155]. Rogel et al. made early investigation on the alkylbenzene-derived amphiphiles on asphaltene surface and explained the positive effect of the size of the ethoxylated chain on the asphaltene stabilization [156].

6.2 Coarse-Grained Level

Due to the unavailability of sufficient computational power, the asphaltene and heavy petroleum complex system cannot be directly simulated at all atomic levels. Dissipative particle dynamics is a coarse-grained particle model that allows the simulation of thousands of compounds on a large time scale. Application of DPD method facilitates the study on asphaltene mesoscale behavior. Zhang et al. proposed a detail method to generate the average asphaltene molecular structure and determine the DPD force parameters [157]. The aggregate structure from DPD is in agreement with MD/MM result and XRD data. The massive number of species in DPD simulation provides information on phase morphology such as oil–water emulsion (Figs. 13 and 14) [157, 158]. Wang et al. lately used DPD to simulate aggregation and diffusion of asphaltene in heptane solvent [159]. Although the DPD has the capacity of modeling a complex system, now the prediction is still qualitative. There are two limitations of asphaltene DPD simulations. First, the detail composition of asphaltene is absent. Thus, there is no accurate composition input for DPD model. Second, accurate force parameters are needed to further develop and give accurate predictions of the physical properties and molecular interactions.

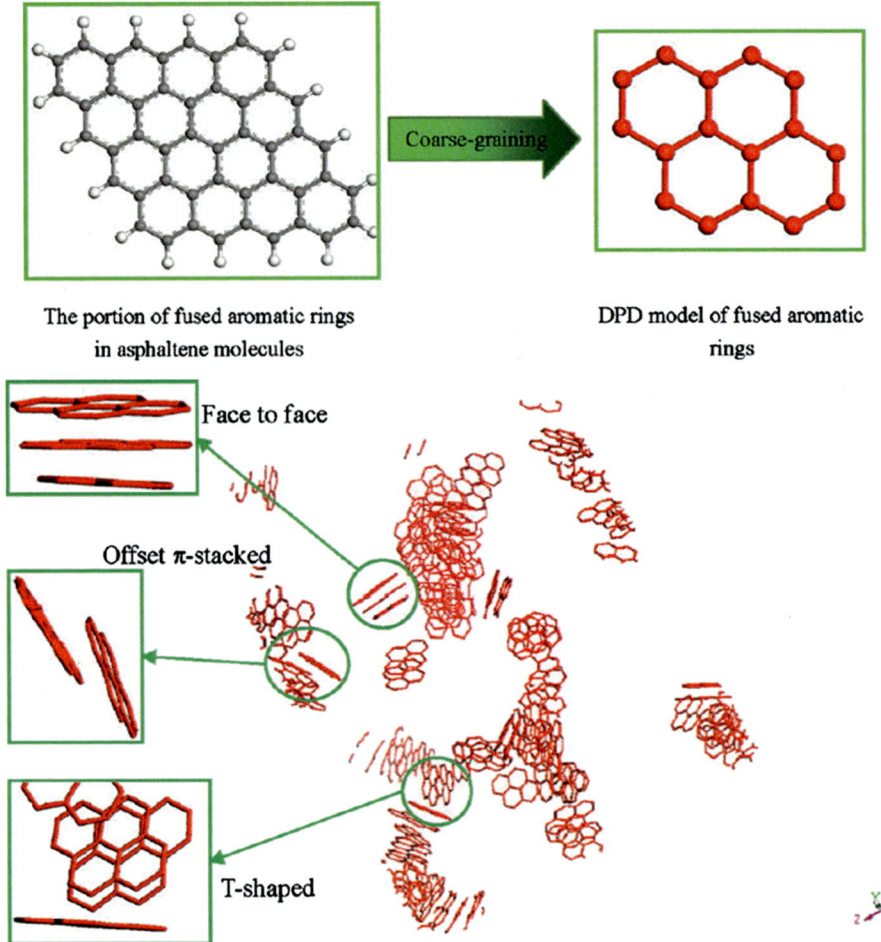

Fig. 13 Coarse-grained model for the fused aromatic rings (*upper*) and morphology of asphaltenes in heavy crude oil (*bottom*) by DPD (reprinted from [157])

7 Thermodynamic Models and Oil Compatibility

Asphaltene precipitation is one of the main reasons for petroleum fouling and compatibility problems. There are many methods available for the measurement of asphaltene participation. The control and prediction of oil compatibility is of considerable interest to the oil industry.

The oil compatibility model provides a simple method for simulation. It is a semi-thermodynamic method proposed by Wiehe and Kennedy that utilized a solubility parameter concept coupled with empirical equations [160]. After measuring the soluble index and insoluble index by several solvent precipitation tests, the compatibility region for oil blends can be calculated. The compatibility of oil

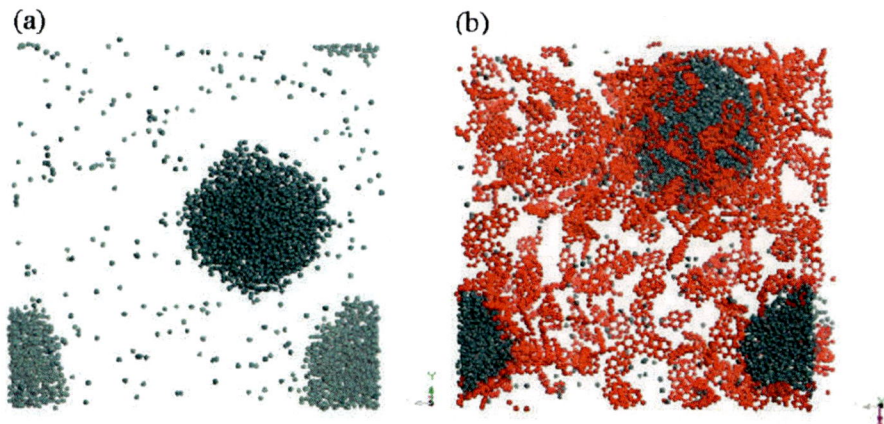

Fig. 14 The equilibrated morphology of heavy crude oil and water emulsion system by DPD. (a) Water; (b) asphaltene and water (reprinted from [157])

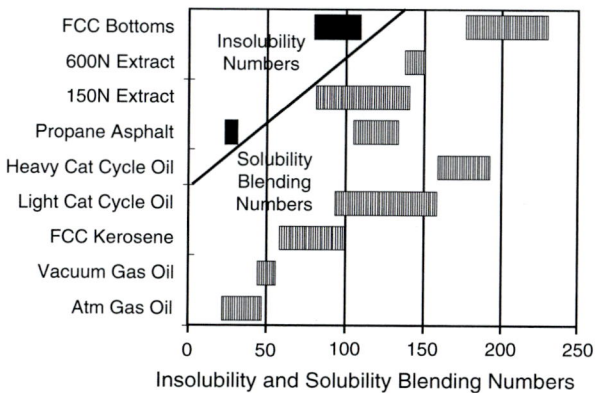

Fig. 15 Range of insolubility numbers and solubility blending numbers for each feed component over days of operation (reprinted from [161])

blends can be calculated from the comparison between the maximum insolubility and mixed solubility number. OCM model was applied to predict asphaltene precipitation condition of oil blend for refinery streams [161]. A sampled compatibility plot of several streams is shown in Fig. 15. The advantage of OCM is its simplicity. The drawback of OCM is the lacking of predicting oil compatibility problem at different conditions. The solubility number and insolubility number is temperature and pressure independent as they are obtained from standard test. For example, the OCM prediction of petroleum fouling is not very accurate [162].

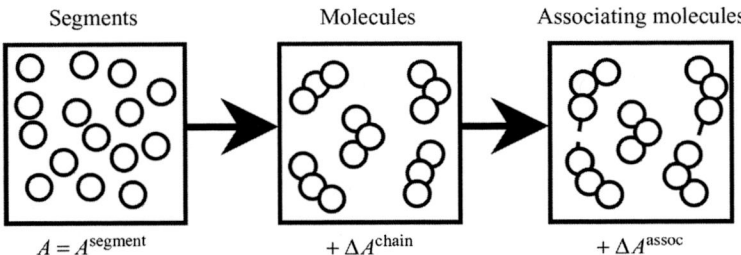

Fig. 16 Contributions to the SAFT equation of state for an associating polyatomic fluid (reprint from [167])

The demand for predicting asphaltene precipitation in reservoirs has promoted the development of a new thermodynamic model. As asphaltene precipitation is caused by molecular interaction and aggregation, the model should account for the formation of asphaltene nanoaggregates. The statistical associating fluid theory equation of state (SAFT EoS) is generally used to solve this problem. The SAFT equation of state was developed by Chapman et al., which extended and simplified Wertheim's theory to characterize the molecular interaction [163–165]. It was then modified by Gross and Sadowski to perturbed-chain SAFT (PC-SAFT) [166]. Different from OCM, SAFT EoS is very mathematically complex. In short, molecules are represented in terms of bonded spherical segment chains. The overall fluid free energy is the sum of free energy of spherical segments, bonding, and directional interaction (Shown in Fig. 16). The PC-SAFT EoS has been successfully applied in petroleum industry to predict the important features of asphaltene precipitation. Ting et al. used PC-SAFT to simulate the phase behavior of an oil with its separator gases [167, 168]. In their approach, petroleum molecules were classified according to SARA composition. The parameters of saturates, aromatics, and resins were derived from an empirical function based on their structure. In general purpose, asphaltene can be considered as single pseudo-component. The asphaltene PC-SAFT EoS parameters in each case were obtained by fitting the titration data. It is notable that the asphaltene MW is determined at a value of ~1,700 Da. Shown in Fig. 17 is the PC-SAFT EoS prediction of oil–gas phase behavior [166]. It can be seen that the asphaltene precipitation onset and the unstable region were calculated with high accuracy. If the structural polydispersity is required, asphaltene can be further fractionated into subfractions by alkanes with different carbon numbers. The success of PC-SAFT EoS is based on the proper illustration of asphaltene aggregation. Although the concept of PC-SAFT EoS is not straightforward to the chemical form of asphaltene nanoaggregate or cluster, its high prediction accuracy reveals that the asphaltene molecular interaction is of significant importance for the phase behavior simulation.

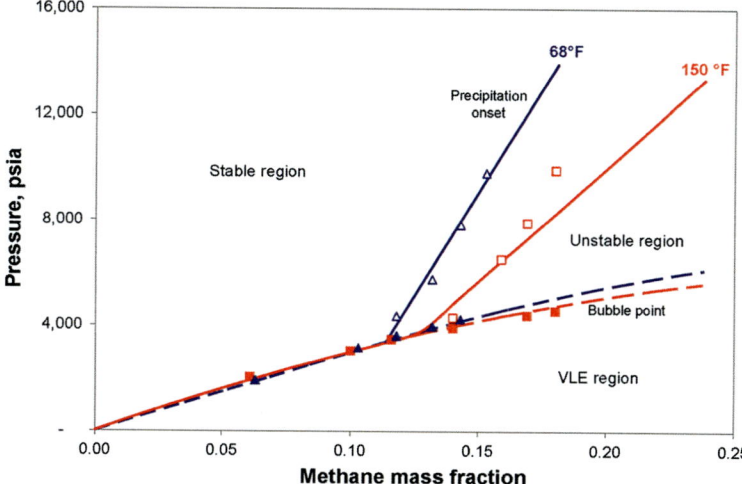

Fig. 17 Experimental bubble points (*filled symbols*) and asphaltene precipitation onsets (*open symbols*) vs. PC-SAFT prediction (*lines*) for a model oil at two different temperatures (reprinted from [166])

8 Concluding Remarks

The petroleum industry has been facing the challenge of organic deposition that seriously affects crude oil extraction and transport. These detrimental processes are mainly caused by the aggregation of asphaltene molecules due to their strong interactions. Motivated by the impact of these phenomena on production and transportation, there are a large number of studies devoted to composition, structure, and aggregation behavior of asphaltenes. It is generally accepted that asphaltene is the heaviest fraction of crude oil, and it is composed of a broad distribution of molecules that have a strong tendency for aggregation. The petroleum asphaltene molecules have small molecular weights (<2000 Da) and are composed of highly fused aromatic rings and short side chains. Both multi-core and single-core motifs exist in asphaltenes. Diverse experimental methods indicate that the size of petroleum asphaltene monomers range from one to three nm, which correspond to roughly seven fused rings. The PAH cores and heteroatoms in asphaltene molecules are the primary sites of interaction, which induce asphaltene molecules to form nanoaggregates with dimensions of 2–10 nm at a concentration of ~100 mg/L. The nanoaggregates can further aggregate, flocculate, and precipitate under certain operating conditions. Asphaltenes behave as a mixture of monomers and of aggregates exhibiting a wide variety of sizes and structures in certain gas or solution conditions.

Up to date, a huge number of studies have been devoted to the investigation of asphaltene molar weight, size, and structure at both molecular and aggregate levels. The controversy about asphaltene molecular weight and size is a direct consequence of the polydispersity of asphaltenes and their tendency to form aggregates at low concentrations. Additional research is still required to improve analytical techniques and experimental methodologies that permit proper determination of molecular composition and structure and which take the tendency of asphaltene aggregate into account. Furthermore, the mechanism of asphaltene aggregation in real oils still remains unresolved, because most scientific papers only focus on one of the levels of asphaltene science. Resulting from the efforts of the petroleum science community, a fuller story from molecule to phase deposition is emerging, but the complete story has not been fully revealed. The research will continue and help to prevent the oil production/processing problems caused by asphaltene aggregation.

Acknowledgments The authors acknowledge the supports by the National Natural Science Foundation of China (NSFC) (no. 21106183, 21476257, U1162204, and U1463207).

References

1. Vafaie-Sefti M, Mousavi-Dehghani SA (2006) Application of association theory to the prediction of asphaltene deposition: deposition due to natural depletion and miscible gas injection processes in petroleum reservoirs. Fluid Phase Equilib 247:182–189
2. Akbarzadeh K, Hammami A, Kharrat A, Zhang D, Allenson S, Creek J, Kabir S, Jamaluddin AJ, Marshall AG, Rodgers RP, Mullins OC, Solbakken T (2007) Asphaltenes-problematic but rich in potential. Oilfield Rev 19:22–43
3. Hayduk W, Buckley WD (1972) Effect of molecular size and shape on diffusivity in dilute liquid solutions. Chem Eng Sci 27:1997–2003
4. Hoshyargar V, Ashrafizadeh SN (2013) Optimization of flow parameters of heavy crude oil-in-water emulsions through pipelines. Ind Eng Chem Res 52:1600–1611
5. Hashmi SM, Quintiliano LA, Firoozabadi A (2010) Polymeric dispersants delay sedimentation in colloidal asphaltene suspensions. Langmuir 26:8021–8029
6. Zhang X, Chodakowski M, Shaw JM (2005) Impact of multiphase behavior on coke deposition in a commercial hydrotreating catalyst under sedimentation conditions. Energy Fuel 19:1405–1411
7. Jian C, Tang T, Bhattacharjee S (2014) Molecular dynamics investigation on the aggregation of Violanthrone78-based model asphaltenes in toluene. Energy Fuel 28:3604–3613
8. Mullins OC (2010) The modified Yen model. Energy Fuel 24:2179–2207
9. Buckley JS (2012) Asphaltene deposition. Energy Fuel 26:4086–4090
10. Akbarzadeh K, Hammami A, Kharrat A, Zhang D, Allenson S, Creek J, Kabir S, Jamaluddin A, Marshall AG, Rodgers RP, Mullins OC, Solbakken T (2007) Asphaltenes-problematic but rich in potential. Oilfield Rev 19:22–43
11. Meeks AC, Goldfarb IJ (1967) Time dependence and drop size effects in the determination of number-average molecular weight by vapor pressure osmometry. Anal Chem 39:908–911
12. Wachter AH, Simon W (1969) Molecular weight determination of polystyrene standards by vapor pressure osmometry. Anal Chem 41:90–94
13. Myers ME, Swarin SJ, Nellis BL (1979) Automated vapor pressure osmometer for determining the molecular weight of polymers. Anal Chem 51:1883–1885
14. Ignasiak T, Strausz OP, Montgomery DS (1977) Oxygen distribution and hydrogen bonding in Athabasca asphaltene. Fuel 56:359–365

15. Yarranton HW, Alboudwarej H, Jakher R (2000) Investigation of asphaltene association with vapor pressure osmometry and interfacial tension measurements. Ind Eng Chem Res 39:2916–2924
16. Agrawala M, Yarranton HW (2001) An asphaltene association model analogous to linear polymerization. Ind Eng Chem Res 40:4664–4672
17. Bressler DC, Wang J, Gawrys KL, Gray MR, Kilpatrick PK, Yarranton HW (2005) Association behavior of pyrene compounds as models for asphaltenes. Energy Fuel 19:1268–1271
18. Yarranton HW, Fox WA, Svrcek WY (2007) Effect of resins on asphaltene self-association and solubility. Can J Chem Eng 85:635–642
19. Zhang L, Shi Q, Zhao C, Zhang N, Chung KH, Xu C, Zhao S (2013) Hindered stepwise aggregation model for molecular weight determination of heavy petroleum fractions by vapor pressure osmometry (VPO). Energy Fuel 27:1331–1336
20. Zhang L, Zhao S, Xu Z, Chung KH, Zhao C, Zhang N, Xu C, Shi Q (2014) Molecular weight and aggregation of heavy petroleum fractions measured by vapor pressure osmometry and a hindered stepwise aggregation model. Energy Fuel 28:6179–6187
21. Trejo F, Ancheyta J, Morgan TJ, Herod AA, Kandiyoti R (2007) Characterization of asphaltenes from hydrotreated products by SEC, LDMS, MALDI, NMR, and XRD. Energy Fuel 21:2121–2128
22. Tanaka R, Sato S, Takanohashi T, Hunt JE, Winans RE (2004) Analysis of the molecular weight distribution of petroleum asphaltenes using laser desorption-mass spectrometry. Energy Fuel 18:1405–1413
23. Pomerantz AE, Hammond MR, Morrow AL, Mullins OC, Zare RN (2008) Two-step laser mass spectrometry of asphaltenes. J Am Chem Soc 130:7216–7217
24. Wu Q, Pomerantz AE, Mullins OC, Zare RN (2013) Laser-based mass spectrometric determination of aggregation numbers for petroleum- and coal-derived asphaltenes. Energy Fuel 28:475–482
25. Wu Q, Pomerantz AE, Mullins OC, Zare RN (2013) Minimization of fragmentation and aggregation by laser desorption laser ionization mass spectrometry. J Am Soc Mass Spectrom 24:1116–1122
26. Qian K, Mennito AS, Edwards KE, Ferrughelli DT (2008) Observation of vanadyl porphyrins and sulfur-containing vanadyl porphyrins in a petroleum asphaltene by atmospheric pressure photoionization Fourier transform ion cyclotron resonance mass spectrometry. Rapid Commun Mass Spectrom 22:2153–2160
27. Qian K, Edwards KE, Siskin M, Olmstead WN, Mennito AS, Dechert GJ, Hoosain NE (2007) Desorption and ionization of heavy petroleum molecules and measurement of molecular weight distributions. Energy Fuel 21:1042–1047
28. Klein GC, Kim S, Rodgers RP, Marshall AG, Yen A, Asomaning S (2006) Mass spectral analysis of asphaltenes. I. Compositional differences between pressure-drop and solvent-drop asphaltenes determined by electrospray ionization Fourier transform ion cyclotron resonance mass spectrometry. Energy Fuel 20:1965–1972
29. McKenna AM, Donald LJ, Fitzsimmons JE, Juyal P, Spicer V, Standing KG, Marshall AG, Rodgers RP (2013) Heavy petroleum composition. 3. Asphaltene aggregation. Energy Fuel 27:1246–1256
30. Peng PA, Fu J, Sheng G, Morales-Izquierdo A, Lown EM, Strausz OP (1999) Ruthenium-ions-catalyzed oxidation of an immature asphaltene: structural features and biomarker distribution. Energy Fuel 13:266–277
31. Peng PA, Morales-Izquierdo A, Hogg A, Strausz OP (1997) Molecular structure of Athabasca asphaltene: sulfide, ether, and ester linkages. Energy Fuel 11:1171–1187
32. Camacho-Bragado GA, Santiago P, Marin-Almazo M, Espinosa M, Romero ET, Murgich J, Rodriguez Lugo V, Lozada-Cassou M, Jose-Yacaman M (2002) Fullerenic structures derived from oil asphaltenes. Carbon 40:2761–2766
33. Sharma A, Groenzin H, Tomita A, Mullins OC (2002) Probing order in asphaltenes and aromatic ring systems by HRTEM. Energy Fuel 16:490–496

34. Sharma A, Mullins O (2007) Insights into molecular and aggregate structures of asphaltenes using HRTEM. Springer, New York, pp 205–229
35. Qian K, Robbins WK, Hughey CA, Cooper HJ, Rodgers RP, Marshall AG (2001) Resolution and identification of elemental compositions for more than 3000 crude acids in heavy petroleum by negative-ion microelectrospray high-field Fourier transform ion cyclotron resonance mass spectrometry. Energy Fuel 15:1505–1511
36. Qian K, Rodgers RP, Hendrickson CL, Emmett MR, Marshall AG (2001) Reading chemical fine print: resolution and identification of 3000 nitrogen-containing aromatic compounds from a single electrospray ionization Fourier transform ion cyclotron resonance mass spectrum of heavy petroleum crude oil. Energy Fuel 15:492–498
37. Marshall AG, Rodgers RP (2003) Petroleomics: the next grand challenge for chemical analysis. Acc Chem Res 37:53–59
38. Zhan D, Fenn JB (2000) Electrospray mass spectrometry of fossil fuels1. Int J Mass Spectrom 194:197–208
39. Qian K, Edwards KE, Diehl JH, Green LA (2004) Fundamentals and applications of electrospray ionization mass spectrometry for petroleum characterization. Energy Fuel 18:1784–1791
40. Sabbah H, Morrow AL, Pomerantz AE, Zare RN (2011) Evidence for island structures as the dominant architecture of asphaltenes. Energy Fuel 25:1597–1604
41. Qian K, Edwards KE, Mennito AS, Freund H, Saeger RB, Hickey KJ, Francisco MA, Yung C, Chawla B, Wu C, Kushnerick JD, Olmstead WN (2012) Determination of structural building blocks in heavy petroleum systems by collision-induced dissociation Fourier transform Ion cyclotron resonance mass spectrometry. Anal Chem 84:4544–4551
42. Podgorski DC, Corilo YE, Nyadong L, Lobodin VV, Bythell BJ, Robbins WK, McKenna AM, Marshall AG, Rodgers RP (2012) Heavy petroleum composition. 5. Compositional and structural continuum of petroleum revealed. Energy Fuel 27:1268–1276
43. Zhang L, Zhang Y, Zhao S, Xu C, Chung K, Shi Q (2013) Characterization of heavy petroleum fraction by positive-ion electrospray ionization FT-ICR mass spectrometry and collision induced dissociation: bond dissociation behavior and aromatic ring architecture of basic nitrogen compounds. Sci China Chem 56:874–882
44. Shattock TR, Arora KK, Vishweshwar P, Zaworotko MJ (2008) Hierarchy of supramolecular synthons: persistent carboxylic acid ··· pyridine hydrogen bonds in cocrystals that also contain a hydroxyl moiety. Cryst Growth Des 8:4533–4545
45. Mohamed S, Tocher DA, Vickers M, Karamertzanis PG, Price SL (2009) Salt or cocrystal? A new series of crystal structures formed from simple pyridines and carboxylic acids. Cryst Growth Des 9:2881–2889
46. Gray MR, Tykwinski RR, Stryker JM, Tan X (2011) Supramolecular assembly model for aggregation of petroleum asphaltenes. Energy Fuel 25:3125–3134
47. Östlund J-A, Nyden M, Auflem IH, Sjoblom J (2003) Interactions between asphaltenes and naphthenic acids. Energy Fuel 17:113–119
48. Tan X, Fenniri H, Gray MR (2009) Water enhances the aggregation of model asphaltenes in solution via hydrogen bonding. Energy Fuel 23:3687–3693
49. Merino-Garcia D, Andersen SI (2004) Interaction of asphaltenes with nonylphenol by microcalorimetry. Langmuir 20:1473–1480
50. Gawrys KL, Blankenship GA, Kilpatrick PK (2006) On the distribution of chemical properties and aggregation of solubility fractions in asphaltenes. Energy Fuel 20:705–714
51. Goual L, Sedghi M, Wang X, Zhu Z (2014) Asphaltene aggregation and impact of alkylphenols. Langmuir 30:5394–5403
52. Da Costa LM, Stoyanov SR, Gusarov S, Tan X, Gray MR, Stryker JM, Tykwinski R, de M. Carneiro JW, Seidl PR, Kovalenko A (2011) Density functional theory investigation of the contributions of π–π stacking and hydrogen-bonding interactions to the aggregation of model asphaltene compounds. Energy Fuel 26(5):2727–2735

53. Stoyanov SR, Yin C-X, Gray MR, Stryker JM, Gusarov S, Kovalenko A (2010) Computational and experimental study of the structure, binding preferences, and spectroscopy of nickel(II) and vanadyl porphyrins in petroleum. J Phys Chem B 114:2180–2188
54. Dickie JP, Yen TF (1967) Macrostructures of the asphaltic fractions by various instrumental methods. Anal Chem 39:1847–1852
55. Kuznicki T, Masliyah JH, Bhattacharjee S (2008) Molecular dynamics study of model molecules resembling asphaltene-like structures in aqueous organic solvent systems. Energy Fuel 22:2379–2389
56. Sedghi M, Goual L, Welch W, Kubelka J (2013) Effect of asphaltene structure on association and aggregation using molecular dynamics. J Phys Chem B 117:5765–5776
57. Castellano O, Gimon R, Soscun H (2011) Theoretical study of the σ–π and π–π interactions in heteroaromatic monocyclic molecular complexes of benzene, pyridine, and thiophene dimers: implications on the resin–asphaltene stability in crude oil. Energy Fuel 25:2526–2541
58. Murgich J (2002) Intermolecular forces in aggregates of asphaltenes and resins. Pet Sci Technol 20:983–997
59. Cockroft SL, Perkins J, Zonta C, Adams H, Spey SE, Low CMR, Vinter JG, Lawson KR, Urch CJ, Hunter CA (2007) Substituent effects on aromatic stacking interactions. Org Biomol Chem 5:1062–1080
60. Prado GHC, de Klerk A (2014) Halogenation of oilsands bitumen, maltenes, and asphaltenes. Energy Fuel 28:4458–4468
61. Yin C-X, Tan X, Müllen K, Stryker JM, Gray MR (2008) Associative π-π interactions of condensed aromatic compounds with vanadyl or nickel porphyrin complexes are not observed in the organic phase. Energy Fuel 22:2465–2469
62. Andersen SI, Jensen JO, Speight JG (2005) X-ray diffraction of subfractions of petroleum asphaltenes. Energy Fuel 19:2371–2377
63. Stachowiak C, Viguié J-R, Grolier J-PE, Rogalski M (2005) Effect of n-alkanes on asphaltene structuring in petroleum oils. Langmuir 21:4824–4829
64. Wang S, Liu J, Zhang L, Masliyah J, Xu Z (2009) Interaction forces between asphaltene surfaces in organic solvents. Langmuir 26:183–190
65. Porte G, Zhou H, Lazzeri V (2002) Reversible description of asphaltene colloidal association and precipitation. Langmuir 19:40–47
66. Mullins OC (2009) The modified Yen model. Energy Fuel 24:2179–2207
67. Mullins OC (2011) The asphaltenes. Annu Rev Anal Chem 4:393–418
68. Mullins OC, Sabbah H, Eyssautier J, Pomerantz AE, Barré L, Andrews AB, Ruiz-Morales Y, Mostowfi F, McFarlane R, Goual L, Lepkowicz R, Cooper T, Orbulescu J, Leblanc RM, Edwards J, Zare RN (2012) Advances in asphaltene science and the Yen–Mullins model. Energy Fuel 26:3986–4003
69. Mullins OC, Pomerantz AE, Zuo JY, Dong C (2014) Downhole fluid analysis and asphaltene science for petroleum reservoir evaluation. Ann Rev Chem Biomol Eng 5:325–345
70. Alboudwarej H, Beck J, Svrcek WY, Yarranton HW, Akbarzadeh K (2002) Sensitivity of asphaltene properties to separation techniques. Energy Fuel 16:462–469
71. Barcenas M, Duda Y (2007) Irreversible colloidal agglomeration in presence of associative inhibitors: computer simulation study. Phys Lett A 365:454–457
72. Barcenas M, Duda Y (2009) Inhibition of irreversible cluster–cluster aggregation of colloids. Colloids Surf A Physicochem Eng Asp 334:137–141
73. Barcenas M, Orea P (2011) Molar-mass distributions of asphaltenes in the presence of inhibitors: experimental and computer calculations. Energy Fuel 25:2100–2108
74. Nellensteyn F (1924) The constitution of asphalt. J Inst Pet Technol 10:311–323
75. Pfeiffer JP, Saal RNJ (1940) Asphaltic bitumen as colloid system. J Phys Chem 44:139–149
76. Sheu EY, De Tar MM, Storm DA, DeCanio SJ (1992) Aggregation and kinetics of asphaltenes in organic solvents. Fuel 71:299–302
77. Andersen SI, Christensen SD (1999) The critical micelle concentration of asphaltenes as measured by calorimetry. Energy Fuel 14:38–42

78. Rogel E, León O, Torres G, Espidel J (2000) Aggregation of asphaltenes in organic solvents using surface tension measurements. Fuel 79:1389–1394
79. Oh K, Ring TA, Deo MD (2004) Asphaltene aggregation in organic solvents. J Colloid Interface Sci 271:212–219
80. Mohamed RS, Ramos ACS, Loh W (1999) Aggregation behavior of two asphaltenic fractions in aromatic solvents. Energy Fuel 13:323–327
81. Oh K, Oblad SC, Hanson FV, Deo MD (2003) Examination of asphaltenes precipitation and self-aggregation. Energy Fuel 17:508–509
82. Merino-Garcia D, Andersen SI (2005) Calorimetric evidence about the application of the concept of CMC to asphaltene self-association. J Dispers Sci Technol 26:217–225
83. Andersen SI, Speight JG (1993) Observations on the critical micelle concentration of asphaltenes. Fuel 72:1343–1344
84. Andreatta G, Bostrom N, Mullins OC (2005) High-Q ultrasonic determination of the critical nanoaggregate concentration of asphaltenes and the critical micelle concentration of standard surfactants. Langmuir 21:2728–2736
85. Zeng H, Song Y-Q, Johnson DL, Mullins OC (2009) Critical nanoaggregate concentration of asphaltenes by direct-current (DC) electrical conductivity. Energy Fuel 23:1201–1208
86. Lisitza NV, Freed DE, Sen PN, Song Y-Q (2009) Study of asphaltene nanoaggregation by nuclear magnetic resonance (NMR). Energy Fuel 23:1189–1193
87. Goual L, Sedghi M, Zeng H, Mostowfi F, McFarlane R, Mullins OC (2011) On the formation and properties of asphaltene nanoaggregates and clusters by DC-conductivity and centrifugation. Fuel 90:2480–2490
88. Mostowfi F, Indo K, Mullins OC, McFarlane R (2009) Asphaltene nanoaggregates studied by centrifugation. Energy Fuel 23:1194–1200
89. Goncalves S, Castillo J, Fernández A, Hung J (2004) Absorbance and fluorescence spectroscopy on the aggregation behavior of asphaltene-toluene solutions. Fuel 83:1823–1828
90. Barre L, Simon S, Palermo T (2008) Solution properties of asphaltenes. Langmuir 24:3709–3717
91. Derakhshesh M, Bergmann A, Gray MR (2012) Occlusion of polyaromatic compounds in asphaltene precipitates suggests porous nanoaggregates. Energy Fuel 27:1748–1751
92. Acevedo S, Cordero T JM, Carrier H, Bouyssiere B, Lobinski R (2009) Trapping of paraffin and other compounds by asphaltenes detected by laser desorption ionization-time of flight mass spectrometry (LDI-TOF MS): role of A1 and A2 asphaltene fractions in this trapping. Energy Fuel 23:842–848
93. Liao Z, Graciaa A, Geng A, Chrostowska A, Creux P (2006) A new low-interference characterization method for hydrocarbons occluded inside asphaltene structures. Appl Geochem 21:833–838
94. Gawrys KL, Blankenship GA, Kilpatrick PK (2006) Solvent entrainment in and flocculation of asphaltenic aggregates probed by small-angle neutron scattering. Langmuir 22:4487–4497
95. Yang C, Liao Z, Zhang L, Creux P (2008) Some biogenic-related compounds occluded inside asphaltene aggregates. Energy Fuel 23:820–827
96. Zhao J, Liao Z, Zhang L, Creux P, Yang C, Chrostowska A, Zhang H, Graciaa A (2010) Comparative studies on compounds occluded inside asphaltenes hierarchically released by increasing amounts of H_2O_2/CH_3COOH. Appl Geochem 25:1330–1338
97. Zielinski L, Saha I, Freed DE, Hürlimann MD, Liu Y (2010) Probing asphaltene aggregation in native crude oils with low-field NMR. Langmuir 26:5014–5021
98. Verruto VJ, Kilpatrick PK (2007) Preferential solvent partitioning within asphaltenic aggregates dissolved in binary solvent mixtures. Energy Fuel 21:1217–1225
99. Savvidis TG, Fenistein D, Barré L, Béhar E (2001) Aggregated structure of flocculated asphaltenes. AICHE J 47:206–211
100. Zajac GW, Sethi NK, Joseph JT, Sellis D, Pareiss C (1977) Maya petroleum asphaltene imaging by scanning tunneling microscopy: verification of structure from 13C and proton nuclear magnetic resonance. Am Chem Soc Div Fuel Chem 42:423–426

101. Groenzin H, Mullins OC (2000) Molecular size and structure of asphaltenes from various sources. Energy Fuel 14:677–684
102. Badre S, Carla Goncalves C, Norinaga K, Gustavson G, Mullins OC (2006) Molecular size and weight of asphaltene and asphaltene solubility fractions from coals, crude oils and bitumen. Fuel 85:1–11
103. Andrews AB, Guerra RE, Mullins OC, Sen PN (2006) Diffusivity of asphaltene molecules by fluorescence correlation spectroscopy. J Phys Chem A 110:8093–8097
104. Kawashima H, Takanohashi T, Iino M, Matsukawa S (2008) Determining asphaltene aggregation in solution from diffusion coefficients as determined by pulsed-field gradient spin – echo 1H NMR. Energy Fuel 22:3989–3993
105. Wargadalam VJ, Norinaga K, Iino M (2002) Size and shape of a coal asphaltene studied by viscosity and diffusion coefficient measurements. Fuel 81:1403–1407
106. Wargadalam VJ, Norinaga K, Iino M (2001) Hydrodynamic properties of coal extracts in pyridine. Energy Fuel 15:1123–1128
107. Nortz RL, Baltus RE, Rahimi P (1990) Determination of the macroscopic structure of heavy oils by measuring hydrodynamic properties. Ind Eng Chem Res 29:1968–1976
108. Sakai M, Sasaki K, Inagaki M (1983) Hydrodynamic studies of dilute pitch solutions: the shape and size of pitch molecules. Carbon 21:593–596
109. Chen Z, Zhao S, Xu Z, Gao J, Xu C (2011) Molecular size and size distribution of petroleum residue. Energy Fuel 25:2109–2114
110. Chen Z, Liu J, Wu Y, Xu Z, Liu X, Zhao S, Xu C (2015) Polydisperse size distribution of monomers and aggregates of sulfur-containing compounds in petroleum residue fractions. Energy Fuel 29(8):4730–4737
111. Eyssautier J, Frot D, Barré L (2012) Structure and dynamic properties of colloidal asphaltene aggregates. Langmuir 28:11997–12004
112. Xu Y, Koga Y, Strausz OP (1995) Characterization of Athabasca asphaltenes by small-angle X-ray scattering. Fuel 74:960–964
113. Tanaka R, Sato E, Hunt JE, Winans RE, Sato S, Takanohashi T (2004) Characterization of asphaltene aggregates using X-ray diffraction and small-angle X-ray scattering. Energy Fuel 18:1118–1125
114. Eyssautier JL, Levitz P, Espinat D, Jestin J, Gummel JRM, Grillo I, Barré LC (2011) Insight into asphaltene nanoaggregate structure inferred by small angle neutron and X-ray scattering. J Phys Chem B 115:6827–6837
115. Amundaraín Hurtado JL, Chodakowski M, Long B, Shaw JM (2011) Characterization of physically and chemically separated Athabasca asphaltenes using small-angle X-ray scattering. Energy Fuel 25:5100–5112
116. Goual L, Abudu A (2009) Predicting the adsorption of asphaltenes from their electrical conductivity. Energy Fuel 24:469–474
117. Goual L (2009) Impedance spectroscopy of petroleum fluids at low frequency. Energy Fuel 23:2090–2094
118. Derakhshesh M, Gray MR, Dechaine GP (2013) Dispersion of asphaltene nanoaggregates and the role of Rayleigh scattering in the absorption of visible electromagnetic radiation by these nanoaggregates. Energy Fuel 27:680–693
119. Dechaine GP, Gray MR (2010) Membrane diffusion measurements do not detect exchange between asphaltene aggregates and solution phase. Energy Fuel 25:509–523
120. Ruiz-Morales Y, Wu X, Mullins OC (2007) Electronic absorption edge of crude oils and asphaltenes analyzed by molecular orbital calculations with optical spectroscopy. Energy Fuel 21:944–952
121. Ruiz-Morales Y, Mullins OC (2008) Measured and simulated electronic absorption and emission spectra of asphaltenes. Energy Fuel 23:1169–1177
122. Bouhadda Y, Bormann D, Sheu E, Bendedouch D, Krallafa A, Daaou M (2007) Characterization of algerian Hassi-Messaoud asphaltene structure using Raman spectrometry and X-ray diffraction. Fuel 86:1855–1864

123. Barrera DM, Ortiz DP, Yarranton HW (2013) Molecular weight and density distributions of asphaltenes from crude oils. Energy Fuel 27:2474–2487
124. Durand E, Clemancey M, Lancelin J-M, Verstraete J, Espinat D, Quoineaud A-A (2009) Aggregation states of asphaltenes: evidence of two chemical behaviors by 1H diffusion-ordered spectroscopy nuclear magnetic resonance. J Phys Chem C 113:16266–16276
125. Durand E, Clemancey M, Lancelin J-M, Verstraete J, Espinat D, Quoineaud A-A (2010) Effect of chemical composition on asphaltenes aggregation. Energy Fuel 24:1051–1062
126. Headen TF, Boek ES, Stellbrink JR, Scheven UM (2009) Small angle neutron scattering (SANS and V-SANS) study of asphaltene aggregates in crude oil. Langmuir 25:422–428
127. Yarranton HW, Ortiz DP, Barrera DM, Baydak EN, Barré L, Frot D, Eyssautier J, Zeng H, Xu Z, Dechaine G, Becerra M, Shaw JM, McKenna AM, Mapolelo MM, Bohne C, Yang Z, Oake J (2013) On the size distribution of self-associated asphaltenes. Energy Fuel 27:5083–5106
128. Korb J-P, Louis-Joseph A, Benamsili L (2013) Probing structure and dynamics of bulk and confined crude oils by multiscale NMR spectroscopy, diffusometry, and relaxometry. J Phys Chem B 117:7002–7014
129. Mullins OC, Seifert DJ, Zuo JY, Zeybek M (2012) Clusters of asphaltene nanoaggregates observed in oilfield reservoirs. Energy Fuel 27:1752–1761
130. Fenistein D, Barré L (2001) Experimental measurement of the mass distribution of petroleum asphaltene aggregates using ultracentrifugation and small-angle X-ray scattering. Fuel 80:283–287
131. Gawrys KL, Kilpatrick PK (2005) Asphaltenic aggregates are polydisperse oblate cylinders. J Colloid Interface Sci 288:325–334
132. Sheu EY, Acevedo S (2001) Effect of pressure and temperature on colloidal structure of furrial crude oil. Energy Fuel 15:702–707
133. Roux J-N, Broseta D, Demé B (2001) SANS study of asphaltene aggregation: concentration and solvent quality effects. Langmuir 17:5085–5092
134. Espinat D, Fenistein D, Barre L, Frot D, Briolant Y (2004) Effects of temperature and pressure on asphaltenes agglomeration in toluene. A light, X-ray, and neutron scattering investigation. Energy Fuel 18:1243–1249
135. Eyssautier J, Hénaut I, Levitz P, Espinat D, Barré L (2011) Organization of asphaltenes in a vacuum residue: a small-angle X-ray scattering (SAXS)–viscosity approach at high temperatures. Energy Fuel 26:2696–2704
136. Acevedo S, García LA, Rodríguez P (2012) Changes of diameter distribution with temperature measured for asphaltenes and their fractions A1 and A2. Impact of these measurements in colloidal and solubility issues of asphaltenes. Energy Fuel 26:1814–1819
137. Fossen M, Kallevik H, Knudsen KD, Sjöblom J (2007) Asphaltenes precipitated by a two-step precipitation procedure. 1. Interfacial tension and solvent properties. Energy Fuel 21:1030–1037
138. León O, Contreras E, Rogel E, Dambakli G, Acevedo S, Carbognani L, Espidel J (2002) Adsorption of native resins on asphaltene particles: a correlation between adsorption and activity. Langmuir 18:5106–5112
139. Kraiwattanawong K, Fogler HS, Gharfeh SG, Singh P, Thomason WH, Chavadej S (2009) Effect of asphaltene dispersants on aggregate size distribution and growth. Energy Fuel 23:1575–1582
140. Pacheco-Sánchez JH, Zaragoza IP, Martínez-Magadán JM (2003) Asphaltene aggregation under vacuum at different temperatures by molecular dynamics. Energy Fuel 17:1346–1355
141. Murgich J, Jesús M, Aray Y (1996) Molecular recognition and molecular mechanics of micelles of some model asphaltenes and resins. Energy Fuel 10:68–76
142. Headen TF, Boek ES, Skipper NT (2009) Evidence for asphaltene nanoaggregation in toluene and heptane from molecular dynamics simulations. Energy Fuel 23:1220–1229

143. Jian C, Tang T, Bhattacharjee S (2013) Probing the effect of side-chain length on the aggregation of a model asphaltene using molecular dynamics simulations. Energy Fuel 27:2057–2067
144. Zhang L, Greenfield ML (2007) Molecular orientation in model asphalts using molecular simulation. Energy Fuel 21:1102–1111
145. Carauta ANM, Seidl PR, Chrisman ECAN, Correia JCG, Menechini PDO, Silva DM, Leal KZ, de Menezes SMC, de Souza WF, Teixeira MAG (2005) Modeling solvent effects on asphaltene dimers. Energy Fuel 19:1245–1251
146. Headen TF, Boek ES (2010) Molecular dynamics simulations of asphaltene aggregation in supercritical carbon dioxide with and without limonene. Energy Fuel 25:503–508
147. Alvarez-Ramirez F, Ramirez-Jaramillo E, Ruiz-Morales Y (2005) Calculation of the interaction potential curve between asphaltene – asphaltene, asphaltene – resin, and resin – resin systems using density functional theory. Energy Fuel 20:195–204
148. Ortega-Rodríguez A, Cruz SA, Gil-Villegas A, Guevara-Rodríguez F, Lira-Galeana C (2003) Molecular view of the asphaltene aggregation behavior in asphaltene – resin mixtures. Energy Fuel 17:1100–1108
149. Moreira da Costa L, Stoyanov SR, Gusarov S, Seidl PR, Walkimar de M. Carneiro J, Kovalenko A (2014) Computational study of the effect of dispersion interactions on the thermochemistry of aggregation of fused polycyclic aromatic hydrocarbons as model asphaltene compounds in solution. J Phys Chem A 118:896–908
150. Yen TF, Chilingarian GV (2000) Asphaltenes and asphalts, vol 2. Elsevier, Amsterdam
151. Rogel E, Carbognani L (2003) Density estimation of asphaltenes using molecular dynamics simulations. Energy Fuel 17:378–386
152. Aray Y, Hernández-Bravo R, Parra JG, Rodríguez J, Coll DS (2011) Exploring the structure–solubility relationship of asphaltene models in toluene, heptane, and amphiphiles using a molecular dynamic atomistic methodology. J Phys Chem A 115:11495–11507
153. Mikami Y, Liang Y, Matsuoka T, Boek ES (2013) Molecular dynamics simulations of asphaltenes at the oil–water interface: from nanoaggregation to thin-film formation. Energy Fuel 27:1838–1845
154. Gao F, Xu Z, Liu G, Yuan S (2014) Molecular dynamics simulation: the behavior of asphaltene in crude oil and at the oil/water interface. Energy Fuel 28:7368–7376
155. Teklebrhan RB, Ge L, Bhattacharjee S, Xu Z, Sjöblom J (2014) Initial partition and aggregation of uncharged polyaromatic molecules at the oil–water interface: a molecular dynamics simulation study. J Phys Chem B 118:1040–1051
156. Rogel E, León O (2001) Study of the adsorption of alkyl-benzene-derived amphiphiles on an asphaltene surface using molecular dynamics simulations. Energy Fuel 15:1077–1086
157. Zhang S-F, Sun LL, Xu J-B, Wu H, Wen H (2010) Aggregate structure in heavy crude oil: using a dissipative particle dynamics based mesoscale platform. Energy Fuel 24:4312–4326
158. Alvarez F, Flores EA, Castro LV, Hernández JG, López A, Vázquez F (2010) Dissipative particle dynamics (DPD) study of crude oil–water emulsions in the presence of a functionalized Co-polymer. Energy Fuel 25:562–567
159. Wang S, Xu J, Wen H (2014) The aggregation and diffusion of asphaltenes studied by GPU-accelerated dissipative particle dynamics. Comput Phys Commun 185:3069–3078
160. Wiehe IA, Kennedy RJ (2000) The oil compatibility model and crude oil incompatibility. Energy Fuel 14:56–59
161. Wiehe IA, Kennedy RJ (1999) Application of the oil compatibility model to refinery streams. Energy Fuel 14:60–63
162. Wiehe IA, Kennedy RJ, Dickakian G (2001) Fouling of nearly incompatible oils. Energy Fuel 15:1057–1058
163. Jackson G, Chapman WG, Gubbins KE (1988) Phase equilibria of associating fluids: spherical molecules with multiple bonding sites. Mol Phys 65:1–31
164. Chapman WG, Gubbins KE, Jackson G, Radosz M (1989) SAFT: equation-of-state solution model for associating fluids. Fluid Phase Equilib 52:31–38

165. Chapman WG, Gubbins KE, Jackson G, Radosz M (1990) New reference equation of state for associating liquids. Ind Eng Chem Res 29:1709–1721
166. Gross J, Sadowski G (2001) Perturbed-chain SAFT: an equation of state based on a perturbation theory for chain molecules. Ind Eng Chem Res 40:1244–1260
167. Ting P, Gonzalez D, Hirasaki G, Chapman W (2007) Application of the PC-SAFT equation of state to asphaltene phase behavior. Springer, New York, pp 301–327
168. Vargas FM, Gonzalez DL, Hirasaki GJ, Chapman WG (2009) Modeling asphaltene phase behavior in crude Oil systems using the perturbed chain form of the statistical associating fluid theory (PC-SAFT) equation of state. Energy Fuel 23:1140–1146

Porphyrins in Heavy Petroleums: A Review

Xu Zhao, Chunming Xu, and Quan Shi

Abstract Vanadium and nickel are the most abundant and troublesome metal compounds present in the organic portions of fossil fuel deposits. These metal compounds may cause significant detrimental impact during refining processes, leading to the deactivation of catalysts used for sulfur and nitrogen removal. Therefore, it is highly desirable to remove vanadium and nickel from petroleum fractions before catalytic hydrogenation and cracking. Vanadium and nickel complexes generally have been classified into porphyrins and non-porphyrins. Studies of the porphyrins have focused extensively on their isolation and identification since their discovery in crude oils and shales. Although it was proposed that non-porphyrins will contain atypical porphyrin or pseudo aromatic tetradentate systems, no non-porphyrin molecules have been identified in crude oil. Ultraviolet–visible (UV–vis) spectroscopy and mass spectrometry are the common analytical techniques used to identify and quantify porphyrins. Due to the high intensity and sensitivity of electronic absorption of UV–vis radiation by porphyrins, approximately half of the vanadium and nickel porphyrins can be identified and quantified by their characteristic UV–vis spectra. The remaining vanadium and nickel compounds, the non-porphyrins, are defined by an absence of distinct UV–vis spectroscopic bands. However, the results of X-ray absorption fine structure (EXAFS) spectroscopy and X-ray absorption near-edge structure (XANES) spectroscopy indicated that these non-porphyrins are indeed still bound in a porphyrinic structure,

---

X. Zhao
State Key Laboratory of Heavy Oil Processing, China University of Petroleum, Beijing 102249, China

CNPC Economics and Technology Research Institute, China National Petroleum Corporation (CNPC), Beijing 100724, China

C. Xu (✉) and Q. Shi
State Key Laboratory of Heavy Oil Processing, China University of Petroleum, Beijing 102249, China
e-mail: xcm@cup.edu.cn

although these metal compounds do not exhibit the characteristic UV–visible absorption. In addition, the recent results of mass spectrometry showed that majority of vanadium and nickel compounds existed in the form of porphyrins, including alkyl porphyrins, sulfur-containing porphyrins, nitrogen-containing porphyrins, and oxygen-containing porphyrins. The porphyrins with O, S, and N atoms should associate more strongly with the asphaltenes than the less polar components. Formation of complexes with other components would shift and attenuate the Soret absorption in UV–visible spectroscopy. The porphyrins are generally believed to chelate or non-covalently associate with aromatic asphaltene components by π–π interactions. Therefore, a portion of the vanadium and nickel compounds give strong optical absorption in the Soret band at ca. 400 nm, and the remainder do not, likely due to formation of complexes or due to chemical modification of the porphyrin ring. The implications of their chemical environments for alternative separation methods are important. In this review article, the current understanding of the forms of vanadium and nickel compounds in heavy petroleum is critically discussed and the methods of separation and demetallization evaluated. Effective separation and ultrahigh mass resolution are needed to resolve these vanadium and nickel compounds. Fourier transform ion cyclotron resonance mass spectrometry (FT-ICR MS), which has the highest available broadband mass resolution, mass resolving power, and mass accuracy, is shown to be a significant method for the qualitative analysis of vanadium and nickel compounds in oil fractions.

Keywords Demetallization • Fourier transform ion cyclotron resonance mass spectrometry (FT-ICR MS) • Nickel porphyrin • Non-porphyrin • Vanadyl porphyrin

Contents

1 Introduction	41
2 Vanadium and Nickel Compounds in Heavy Petroleums	42
3 Isolation of Porphyrins from Heavy Petroleums	44
3.1 Solvent Extraction for Primary Enrichment Separation	44
3.2 Column Chromatography for Porphyrins Separation	45
3.3 Purification Methods for the Specific Porphyrins	45
4 Chemical Characterization of Porphyrins	46
4.1 Identification and Quantification by UV–Visible Spectroscopy	47
4.2 Molecular Characterization by Mass Spectrometry	50
4.3 Structure Characterization by X-Ray Absorption Spectroscopy	57
4.4 Other Methods	58
5 Porphyrins as a Maturity Parameter for Petroleum Thermal Evolution	60
6 Demetallization Technologies for Petroleum Upgrading	61
7 Summary and Future Prospects	63
References	63

1 Introduction

Heavy petroleum and natural bitumen are oils set apart by their high viscosity (resistance to flow) and high density (low API gravity) [1]. The reserves of heavy petroleum are 8.90 trillion barrels, much larger than the 1.64 trillion barrels of conventional crude oil [1, 2].

Vanadium and nickel are the major metal elements which occur in crude oils in appreciable quantities as soluble organic complexes. It is well known that vanadium and nickel are present in the heavy petroleum in amounts ranging from 1 to 103 ppm depending on the oil source. Vanadium and nickel are the most abundant and troublesome metal compounds present in the organic portions of fossil fuel deposits. These metal compounds could cause significant detrimental impact during upgrading processes and often cause serious problems in catalytic cracking and hydrodesulfurization units. Pore blocking, fouling of active sites, and change in catalyst selectivity are typical problems [3] which could be reduced by eliminating these problematic compounds before further upgrading. Therefore, better understanding of their amount, distribution, physicochemical properties, and chemical environment is useful in the development of catalytic processes and demetallization.

The majority of vanadium and nickel compounds are present in heavy petroleum as porphyrins. Porphyrins have been extensively studied since their discovery in crude oils and shales as "molecular fossils" by Treibs [4, 5]. The UV–vis spectroscopy is a commonly used analytical technique for identification and quantification of porphyrins, because of its high intensity and sensitivity for porphyrins and metal-free porphyrins. A portion of the vanadium and nickel compounds give strong optical absorption in the Soret band at ca. 400 nm; however, because of the complex nature of the hydrocarbon mixture, it is difficult to characterize vanadium and nickel compounds present in low concentrations. This is probably due to the formation of complexes or the chemical modification of the porphyrin ring [6]. Hence, non-porphyrins are defined by an absence of distinct UV–vis spectroscopic bands. They could contain atypical porphyrin or pseudo aromatic tetradentate ligand systems; however, no non-porphyrins were structurally characterized from heavy petroleum. The isolation of porphyrins from crude oils could reduce or eliminate interferences from other petroleum compounds which enable molecular characterization by ultrahigh mass resolution mass spectrometry.

Mass spectrometry is the most useful technique for the characterization of porphyrins, which provides molecular weights and elemental composition of porphyrin molecules. Several series of porphyrins have been identified in fossil fuels using mass spectrometry; porphyrins contain a core structure of N_4VO. Among them, etioporphyrins (ETIO) and deoxophylloerythroetio porphyrins (DPEP) are the two most common porphyrins found in petroleum. These compounds have double bond equivalence (DBE) of 17 and 18, respectively [7]. Other petroporphyrins include dicyclic-deoxophylloerythroetio porphyrins (Di-DPEP), rhodo-etioporphyrins (rhodo-ETIO), rhodo-deoxophylloerythroetio porphyrins

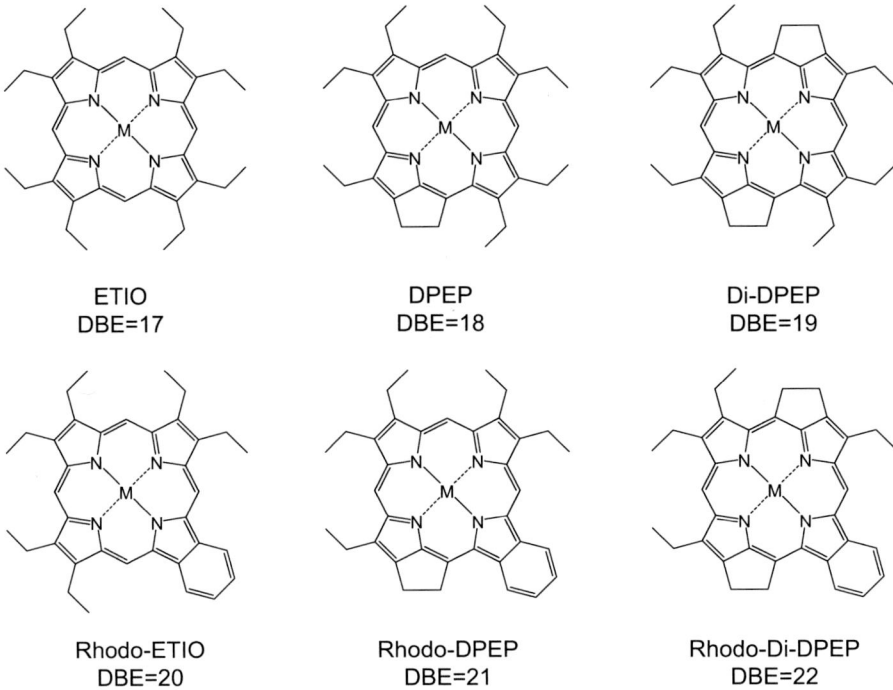

Fig. 1 Common porphyrins in heavy petroleum

(rhodo-DPEP), and rhodo-dicyclic-deoxophylloerythroetio (rhodo-Di-DPEP). Structures of these compounds were shown in Fig. 1.

This chapter reviews the literature on vanadium and nickel compounds in heavy petroleum fractions. The current understanding of the forms of vanadium and nickel compounds in heavy petroleum is critically discussed, including alkyl porphyrins, sulfur-containing porphyrins, nitrogen-containing porphyrins, and oxygen-containing porphyrins. It is widely believed that a better understanding of the chemical environment of the metals in heavy petroleum, amount, distribution, and their chemical properties should be helpful for the development of more efficient demetallization and catalytic processes. In recent years, much effort has been made to develop demetallization technology, which is also discussed in this article.

2 Vanadium and Nickel Compounds in Heavy Petroleums

Nearly one half of the elements in the periodic table have been identified as trace elements in the petroleum. Vanadium and nickel are the two most abundant and problematic metals in petroleum with vanadium commonly reaching concentrations up to 1,200 ppm and nickel up to 150 ppm depending on the origin of the petroleum [8].

Previous studies have shown that there is a systematic variation in the nickel and vanadium content of petroleums which can be related to the rock type source and depositional environment. Many authors have used nickel/vanadium ratios for empirical oil–oil and oil–source correlation, but few attempts have been made to rationalize the processes responsible for variation of metal concentration in oils. Lewan [9, 10] has shown that the rock type source and depositional environment have a profound effect on the predicted levels of nickel and vanadium in source rocks. Low nickel/vanadium ratios (<0.5) are expected for these kinds of organic matter where vanadium incorporation into porphyrins is expected to be favorable compared to nickel incorporation, such as the petroleums from Abu Dhabi and Gulf of Suez. However, the petroleums from China, Indonesia, Gippsland Basin, and Australia have higher nickel/vanadium ratios, and both the input and preservation of porphyrins are expected to be low for this organic matter. Overall, both the absolute quantity of metals and their relative content are predicted to be class-type dependent. Barwise [11] pointed out the highest concentrations of metal are to be found in low-maturity crude oils derived from source rocks which have a low clay content and a high organic sulfur content such as carbonate source rocks; and moderate quantities of metals are found in oils derived from marine shales or lacustrine source rocks, whereas little nickel or vanadium is found in land plant-derived oils. In addition, the content of vanadium is generally much higher than nickel in marine oil; the majority of Chinese petroleums belong to the continental oils, which have higher nickel contents than vanadium. However, Tahe heavy petroleum has much higher vanadium concentration than nickel [12], which is located in the Tarim Basin, a marine Paleozoic oilfield in China. To sum up, the metal content of petroleums can provide a useful insight into the source of a petroleum accumulation, especially when combined with classification parameters derived from molecular, isotopic, and bulk parameters.

Vanadium and nickel compounds have a diverse distribution in various boiling point fractions of petroleum. Barwise et al. [13] found that majority of vanadium is enriched in residue fractions (>500°C) and only a little vanadium is contained in the distillates (350–500°C). Reynolds [14] separated some atmospheric residue to SARA (saturate, aromatic, resin, and asphaltene) and found that majority of vanadium and nickel were in the polar fractions, both resin and asphaltene. Zhang et al. [15] determined the distribution of vanadium and nickel in the Venezuela Orinoco petroleum vacuum residue and its supercritical fluid extraction subfractions; the results showed that most of vanadium and nickel were enriched in the end cuts containing a lot of resin and asphaltene. Pearson et al. [16] investigated the distribution of vanadium and nickel in ABN fractions (acid, base, and neutral fractions); they found that these metal compounds were present in all of the ABN fractions.

The researchers found that majority of the vanadium and nickel compounds are contained within the highly aromatic and highly polar resin and asphaltene fractions. In such fractions, compounds associate and/or aggregate significantly, which pose some significant difficulties in removing the metal compounds. As discussed later, the metals are known to be bound into petroleum partly as organometallic complexes (porphyrins) and partly as high molecular weight complexes ("nonporphyrins") associated with asphaltenic components of crude oil. Better understanding of the distribution of metals is significant and useful for removing and/or demetallization.

3 Isolation of Porphyrins from Heavy Petroleums

As discussed above, owing to the low concentration of porphyrins, it is necessary to separate them from petroleum before they can be analyzed by mass spectrometry. The strategies used for the isolation of porphyrins depend both on the nature of the sample containing the porphyrins, and on the type of studies to be undertaken on the extract. If the study of total porphyrin mixtures is the objective, emphasis must be placed on optimizing the recovery of the porphyrins. Conversely, if characterization of individual components is the goal, the emphasis lies in obtaining samples of the highest possible purity.

Numerous separation approaches have been reported, and most of them usually involve these steps: solvent extraction, chromatography, acid demetallization, and purification. The initial extraction procedures are largely independent of the subsequent studies undertaken. The purification of total porphyrin mixture and the purification techniques for the study of individual porphyrins are discussed in turn.

3.1 Solvent Extraction for Primary Enrichment Separation

There are significantly different problems in working with bitumens and crude oils compared to working with sediments. As for the sediments, the samples must be pulverized for efficient extraction. The porphyrin concentrate is usually isolated either by Soxhlet extraction, by sonication, or by means of a ball mill, which has been used extensively by Baker [17]. The technique of ball milling with an appropriate solvent is probably the best method as it avoids the risks of thermal alteration and bonds cleavage encountered in Soxhlet extraction and sonication, respectively.

However, as for the bitumens and petroleums, these substances are almost exclusively organic and usually viscous; the extraction techniques describe above are generally inappropriate. Some authors still used solvent extraction to isolate porphyrins. In this approach, the oil and bitumen are dispersed in a solid surface, e.g., alumina [17], cellulose [18], siliceous earth [19], diatomite [12], or silica gel [20], and extracted initially with either methanol, toluene–methanol, or acetone–methanol. Generally, solvent extraction is the initial extraction procedure, followed by column chromatography to obtain purified total porphyrins.

3.2 Column Chromatography for Porphyrins Separation

Actually, in addition to the solvent extraction, the porphyrins may be isolated by column chromatography of the entire petroleum. This method has the advantage that the generation of artifacts is avoided [21]. Column chromatography using alumina or silica gel is by far the most common purification technique. Vanadyl porphyrins are likely to have higher polarities than nickel porphyrins because of the presence of the VO group. As a result, the former are usually eluted following the latter when separated on the silica gel column. The nickel porphyrins could be separated from the vanadyl porphyrins on the alumina or silica gel (100–200 mesh) column by gradient elution.

Alumina column chromatography separations are developed by gradient elution using a variety of solvent mixtures and are monitored by visible spectroscopy [22–28]. A gradient elution from hexane to dichloromethane via toluene is an effective system. The nickel porphyrins are likely to elute in 30–50% toluene in hexane and the vanadyl porphyrins in 30–60% dichloromethane in toluene. In addition, a wide range of solvent gradients have been used in silica gel chromatography [17, 21, 22, 24, 25, 29, 30]. A gradient elution of a silica gel column (100–200 mesh) from hexane to chloroform is usefully a very effective system. Most of the nickel porphyrins should be eluted by 10–20% $CHCl_3$ in hexane. The vanadyl porphyrins should be eluted by 30–50% $CHCl_3$ in hexane. Generally it is advisable to purify on both alumina and silica columns.

3.3 Purification Methods for the Specific Porphyrins

The lack of effective extraction procedures, which selectively remove the porphyrins from the petroleum matrix, hampers their determination. As discussed above, solvent extraction and column chromatography could be used to concentrate the porphyrins, but it could not meet the resolution requirements of the traditional analytical instruments. Using strong acids such as methane sulfonic acid to purify the porphyrins mixtures is the most popular method historically and prepares the metal-free porphyrins for chromatographic characterization [31]. Therefore, the traditional analytical instruments such as mass spectrometry at low ionizing voltages could be used to determine the type and distribution of metal-free porphyrins. Although it has been confirmed that the porphyrin structures are unchanged during MSA demetallization [32], it has some drawbacks. The data on the porphyrin distributions is lost; the best yield of porphyrins claimed is usually about 70%. In addition, it is assumed that all porphyrin types are extracted equally well which is untrue since vanadyl rhodo porphyrins are reported to be stable in sulfuric acid [33].

Thin-layer chromatography (TLC) has been used and useful in the final stage of purification. Vanadium and nickel porphyrins have been purified on silica gel plates, heptane–tetrahydrofuran (5:1, vol:vol), and heptane–dichloromethane (7:3, vol:vol) as eluant respectively [34]. However, Ail et al. [28] found that the nickel porphyrins decomposed on the silica gel plate when they tried to purify nickel porphyrins using TLC. Chakraborty [35] also noticed this phenomenon and attributed it to the lability of nickel porphyrins compared to the vanadyl porphyrins.

Another purification method which was always useful and popular is high-performance liquid chromatography (HPLC) [18, 36, 37]. HPLC has good resolution, and it is very sensitive to porphyrins when a visible detector is used at about 410 nm. However, HajIbrahim [38] has pointed out that HPLC has two shortcomings: incomplete separation of the components and the inability to resolve positional isomers. It is possible that more than one phase of LC, like silica columns and/or octadecylsilane (ODS) columns et al., are needed when analyzing complex mixtures such as petroporphyrins. Based on this research work, Hohn [39], Sundararaman [40], Quirke [41], and Peng [42], respectively, obtained the molecular porphyrins data from HPLC traces to investigate the maturation and migration of petroleum. With better chromatography resolution and structural identification of porphyrins, more detailed information about relationships among oils and the nature of geochemical processes should be forthcoming.

4 Chemical Characterization of Porphyrins

Oxovanadium (IV) and nickel (II) complexes of alkyl porphyrins are widely distributed in petroleum, oil shales, and maturing sedimentary bitumen. They have been extensively studied, and attention has focused on their isolation and identification since their discovery in crude oils and shales by Treibs [4, 5]. He developed a detailed scheme and postulated that porphyrins were formed from chlorophylls and hames in the organisms, as can be seen in Figs. 2 and 3. Petroporphyrins are of special significance in theoretical research of petroleum genesis. Vanadium and nickel contents of carbonaceous sedimentary rocks can be used as a guide for petroleum exploration, for the evaluation of ancient depositional environments, and for the estimation of vanadium and nickel resources.

The structures shown in Fig. 1 indicate that ETIO and DPEP are the principal porphyrins in petroleum and the mostly used parameter is the $\sum$DPEP/$\sum$ETIO ratio. The ratio has been widely used as a maturity parameter for crude oil and source rocks [17, 43–47]. On the other hand, it should be noted that the type and the molecular weight distribution of porphyrins may have impacts on their hydrodemetallation (HDM) reaction behaviors.

The separation and purification of porphyrins and "non-porphyrins" in heavy petroleum and the recognition of the type, distribution, molecular weight, and molecular structure characteristics of vanadium and nickel compounds at the molecular level could enrich the scientific understanding of porphyrins compounds. These results could have an important consequence for improving the level of applied basic theory on geochemistry and upgrading processes. A wide variety of

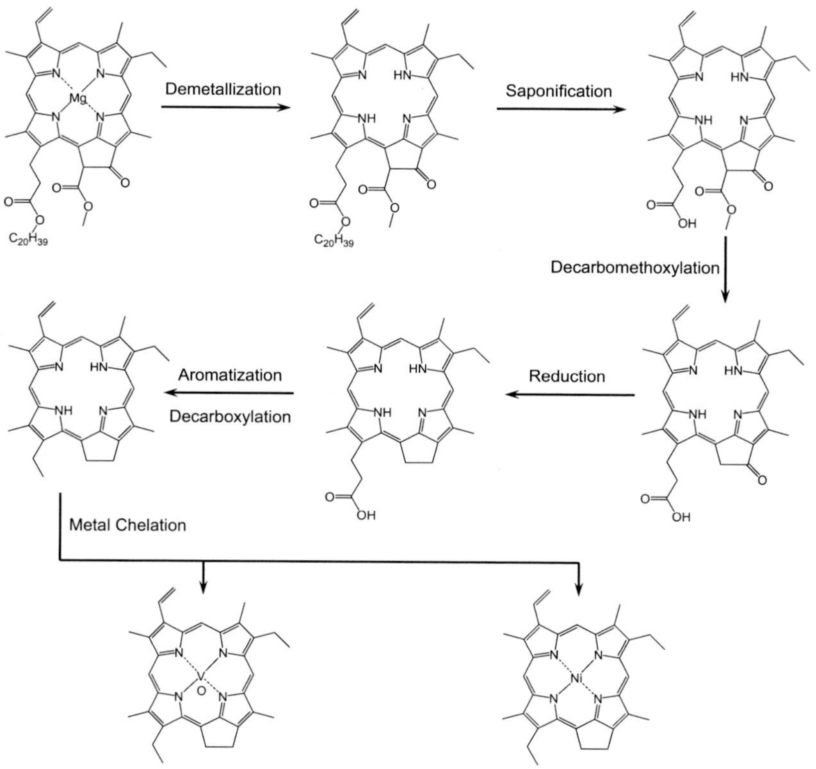

Fig. 2 Proposed evolution process of chlorophyll

analytical methods have been employed to determine the exact molecular form of the vanadium and nickel compounds.

4.1 Identification and Quantification by UV–Visible Spectroscopy

The UV–vis spectroscopy is a commonly used analytical technique for identification and quantification of porphyrins, because of its high intensity and sensitivity of porphyrins and metal-free porphyrins to electronic absorption of UV–vis radiation. For porphyrins, the most intense absorption occurs in the vicinity of 400 nm and is termed the Soret band, which is near the UV violet. There are two characteristic bands in the visible region between 500 and 600 nm, the location of which is dependent on the peripheral substitution of the macrocycle, referring to α and β bands. In addition, the concentration of porphyrins can be determined by the absorption intensities of these bands. Due to the high intensity and sensitivity of electronic absorption of UV–vis radiation by porphyrins, approximately half of the

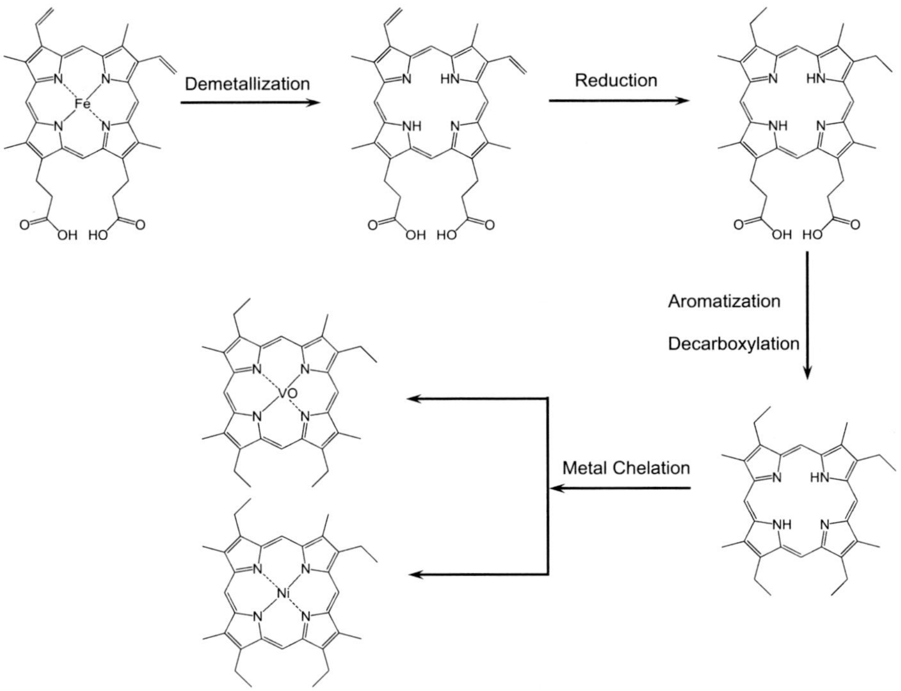

Fig. 3 Proposed evolution process of heme

vanadium and nickel porphyrins can be identified and quantified by their characteristic UV–vis spectra. The remaining vanadium and nickel compounds, the non-porphyrins, are defined by an absence of distinct UV–vis spectroscopic bands.

Since Treibs [4, 5] demonstrated the existence of a complex of vanadium and a porphyrin in various bituminous substances found in Europe, the absorption spectrum is commonly used. Glebovskaya and Vol'kenshstein [48] investigated bituminous clays and Russian petroleums and found evidences of vanadium porphyrin complexes in many of their samples. The studies and conclusions advanced by these European investigators concerning the chemical structures of the vanadium complexes were based upon absorption spectral findings. Skinner [49] determined the physical and chemical state of vanadium in Santa Maria Valley Crude Oil and the properties of the vanadium bearer. It was suggested that information be gathered on the origin of the vanadium bearer and the manner in which it was introduced into the crude. These considerations are not only of academic interest, but have a definite practical aspect as well.

The discovery of the porphyrins in petroleum is a significant milestone in the study of the origin of petroleum. Many investigators used absorption spectral data to determine the existence of porphyrins in petroleum, even when mass spectrometry became available in the 1960s [50, 51]. However, there are several factors which can have an impact on the measurement of UV–vis spectroscopy, such as the

Table 1 Characteristic absorption bands of different vanadyl porphyrins in dichloromethane [53]

Vanadyl porphyrins	Wavelength (nm)		
	Soret band	β band	α band
No substitution	399.4	523.8	559.4
ETIO	406.6	532.8	570.7
Octaethyl	407.3	533.2	570.9
DPEP	410.5	533.3	573.0
Benzo	414.0	544.7	578.7

solvent system, coordination, association, and peripheral substituents of porphyrins. Freeman et al. [52] studied the absorbance of VOOEP in methylene chloride, chloroform, 1–2 dichloroethane, ethyl acetate, and toluene. They obtained the result that the solvent had an impact on the extinction coefficient of porphyrins. In addition, Freeman et al. [53] identified the peak locations for various vanadyl porphyrins in dichloromethane solvent using third derivative UV–visible spectroscopy, and some of the results are shown in Table 1, which indicated that the peripheral substitution of the porphyrin macrocycle can have an impact on the location of the UV–visible peaks.

It was also noted that the peripheral substituent has an impact on the magnitude of absorption peaks. The extinction coefficient of VO-DPEP is about 4 times lower than VO-ETIO, which were measured by Foster [54] and Cantú [55]. If the extinction coefficient of VO–ETIO is used for quantification of porphyrin mixtures, the concentration result of vanadyl porphyrins will be underestimated. That is one reason why approximately half of the vanadium and nickel porphyrins can be identified and quantified by their characteristic UV–vis spectra. Another important reason is that the occurrence of coordination or dimerization/aggregation can alter the UV–visible spectra of porphyrins. The majority of porphyrins were present with residue fractions containing the polar fractions; both resin and asphaltene, the quantitation calculations, were based on the extinction coefficient of isolated porphyrins in solution.

Although several factors can have an impact on the measurement, UV–vis spectroscopy was indispensable and invaluable for the characterization of porphyrins. Furthermore, a number of authors have attempted to determine the type of porphyrins by measuring the absorption peaks of free base porphyrins. After removal of the central Ni^{2+} or VO^{2+}, the corresponding free base porphyrins have four visible peaks, which are labeled as I, II, III, and IV peaks from long wavelength to short wavelength, respectively, and the relative intensities of the peaks are useful in assigning porphyrin types [17, 23, 32, 47]. The ETIO-type visible spectrum of free base porphyrins is characterized by peak intensities in the order IV > III > II > I. The order becomes IV > I > II > III for the DPEP type, while for the benzo type (benzo series and benzo-DPEP series), the intensity order is III > IV > II > I [17].

As for the quantification of porphyrins, the majority of authors attempted to use the extinction coefficients of isolated porphyrins or model compounds; Freeman [56] applied derivative UV–visible spectroscopy to quantify the concentration of porphyrins in deasphalted bitumen samples. They found the significant absorbance by polycyclic aromatic hydrocarbons (PAHs) compounds in the region of the Soret band; therefore, they had to use the α band. In order to counteract the reduced sensitivity of this peak, data smoothing and second derivative algorithms were applied to the spectra. These data analysis algorithms serve to improve the signal-to-noise ratio, which is a major difficulty when dealing with complex mixtures where significant background absorbance is present. In the end, they concluded that the optimal algorithm was a second derivative, three-point sliding average algorithm. Freeman et al. [53] extended this analysis further by applying third derivative UV–visible spectroscopy for the qualitative identification of porphyrins. The use of the third derivative of the absorbance allows for a much more precise identification of the exact wavelength (to within ±0.1 nm) of an absorbance maximum since the third derivative is characterized by a steep zero crossing at an absorbance peak. This method allowed the investigators to differentiate a number of different porphyrins on the basis of the UV–visible spectra alone. Unfortunately, this method requires that the porphyrins be separated and/or purified prior to analysis. The complex mixtures in vacuum residue cannot be analyzed directly to identify the different porphyrins because of significant spectral interferences.

4.2 Molecular Characterization by Mass Spectrometry

In early studies of porphyrins in the geosphere, the complexity of these biomarkers was not understood. Following Treibs' discoveries, many workers have broadened and attended to confirm Treibs' ideas as to the source and method of formation of the porphyrins, and there have been scattered indications that the porphyrins in petroleum were not a single species but rather a series of compounds, some with considerably higher molecular weights. Therefore, mass spectrometry methods for porphyrin characterization were developed in a comprehensive manner since 1960 [50, 51].

Mass spectrometry has provided molecular weights and empirical formulae for the porphyrins [57]. Electron impact ionization (EI) mass spectrometry (MS) has been the most widely used technique in this field. However, the bulk of early work was conducted on demetallated fractions; demetallization could make porphyrins sufficiently volatile to be analyzed. Many authors [23, 32, 58–63] have used EI MS to characterize metal-free porphyrins which were extracted by MSA. They proved the existence of ETIO and DPEP and found rhodo-ETIO, rhodo-DPEP, Di-DPEP, and rhodo-Di-DPEP type of porphyrins. The molecular weights of metal-free porphyrins are related to their types, which have been well established [17, 23]. Metal-free porphyrins of the ETIO series have molecular weights of $310+14n$, where n is an integer. Likewise, metal-free porphyrins of the DPEP series have molecular weights of $308+14m$. Such results allow the determination of the porphyrin homologous series, the main series ratio ($\sum$DPEP/$\sum$ETIO), the carbon number range, and the maximum peak for each series.

For EI MS, to reduce the β cleavage of the peripheral constituents and obtain the molecular ions only, the ionizing voltage is often reduced from 70 to 10–20 eV. But this inevitably reduces its sensitivity [64]. In addition, although separation and purification were performed on porphyrins, some petroleum matrix or contaminants could be present in the purified fraction, which would undermine the EI MS characterization of the porphyrins.

Because porphyrins concentrated in more complex and high-boiling heavy crude oils, high resolution is necessary to distinguish chemically different components. Therefore, chemical ionization (CI) mass spectrometry [65], high-resolution and low-energy electron ionization (EI) mass spectrometry [66], magnetic sector mass spectrometry [67], time-of-flight (TOF) mass spectrometry [47], supercritical fluid chromatography/mass spectrometry (SFC/MS) [68, 69], gas chromatography/mass spectrometry (GC/MS) [13, 69], and liquid chromatography/mass spectrometry (LC/MS) [70], as well as tandem MS (MS/MS) [71, 72], have been used to characterize porphyrins at the molecular level.

In addition, Van Berkel et al. [73] firstly used electrospray ionization mass spectrometry (ESI MS) to detect porphyrins. Electrospray ionization (ESI) source is a kind of soft ionization technique; usually the ionization makes molecular ion peak, instead of the fragment ion peaks [74, 75]. ESI cannot directly ionize hydrocarbon compounds in petroleums, but can ionize the polar compounds [76–80], such as porphyrins.

Over the past decade, mass spectrometry has been revolutionized by access to instruments of increasingly high mass-resolving power. For small molecules up to ~400 Da, it is possible to determine elemental compositions of thousands of chemical components simultaneously from accurate mass measurements. The history of spectroscopy is the history of resolution [81]. Fourier transform ion cyclotron resonance mass spectrometry (FT-ICR MS) has the highest available broadband mass resolution, mass resolving power, and mass accuracy, which enables the assignment of a unique elemental composition to each peak in the mass spectrum [82, 83].

Rodgers et al. [84] first used ultrahigh-resolution electrospray ionization Fourier transform ion cyclotron resonance mass spectrometry (ESI FT-ICR MS) to characterize vanadium porphyrins isolated from Cerro Negro heavy crude, without metallization by MSA. They identified homologues of vanadyl porphyrins and determined the type and distribution at the molecular level. This work provides a valuable reference base for direct speciation of porphyrins in petroleum and also provides an access to a much higher level of compositional and structural detail in characterizing porphyrins.

As discussed above, ESI can just ionize the polar compounds in petroleums. However, atmospheric pressure photoionization (APPI) FT-ICR MS could be used to supplement the ESI ionization source; the ionization ranges from polar compound extended to weak polar and nonpolar compounds. APPI can ionize the aromatic hydrocarbons, sulfur compounds, nitrogen compounds, and other in the low polar compounds. Purcell et al. [85] firstly couple the APPI to FT-ICR MS for characterization of polar and nonpolar species in petroleums. Qian et al. [86]

analyzed a whole asphaltene sample using APPI FT-ICR MS. They not only identify several expected forms of vanadyl porphyrins (VO-ETIO, VO-DPEP, and VO-rhodo) but also identified several sulfur-containing vanadyl porphyrins. This report is the first evidence of sulfur species directly attached to vanadyl porphyrins.

Generally, the porphyrins samples analyzed with mass spectrometry are chemical extracts of the petroleum rather than the whole sample or even an asphaltene sample. However, McKenna et al. [7] firstly detected and identify vanadyl porphyrins in a South American heavy petroleum and an Athabasca bitumen asphaltene without prior fractionation or sample treatment by APPI FTICR MS. They detected the vanadyl porphyrins by the DBE number, which provides the high mass-resolving power and accurate mass to unambiguously assign elemental compositions to each peak in the mass spectrum.

Qian et al. [87] firstly detected and identified nickel porphyrins in a petroleum fraction by APPI FT-ICR MS. They pointed out that difficulty in nickel porphyrin detection was due to the low analytical sensitivity and the interferences of very close mass overlap which cannot be resolved by current FT-ICR MS. Enrichment of nickel compounds and low sulfur content are the two critical factors to the identification of nickel porphyrins.

Based on the early research work, Zhao et al. [20, 88] used the ESI FT-ICR MS to detect vanadium compounds in Venezuela Orinoco heavy petroleum. Venezuela Orinoco heavy crude petroleum was subjected to solvent extraction and silica gel chromatographic separation and sequentially separated into several subfractions to determine the contents and types of vanadyl porphyrins contained in the products. Vanadium contents in each subfraction were detected using an atomic absorption spectrometer (AAS) combined with the characterization of vanadyl porphyrins by ultraviolet–visible (UV–vis) spectroscopy and positive ion electrospray ionization (ESI) Fourier transform ion cyclotron resonance mass spectrometry (FT-ICR MS). Six types of petroleum vanadyl porphyrins, which have been identified previously, were well characterized with the detailed fractionation. In addition, 29 new series of vanadyl porphyrins, totally 370 compounds with distinct molecular composition, were also detected. These compounds probably have the main structure of porphyrin, with the addition of more aromatic rings, carbonyl and/or carboxyl, thiophene, and amino functional groups at the periphery of the porphyrin structure, corresponding to molecular series below:

1. $C_nH_{2n-40}N_4V_1O_1$ ($36 \leq n \leq 58$), $C_nH_{2n-42}N_4V_1O_1$ ($37 \leq n \leq 57$), $C_nH_{2n-44}N_4V_1O_1$ ($38 \leq n \leq 59$), $C_nH_{2n-46}N_4V_1O_1$ ($43 \leq n \leq 54$), $C_nH_{2n-48}N_4V_1O_1$ ($45 \leq n \leq 55$)
2. $C_nH_{2n-30}N_4V_1O_2$ ($27 \leq n \leq 43$), $C_nH_{2n-32}N_4V_1O_2$ ($28 \leq n \leq 42$), $C_nH_{2n-34}N_4V_1O_2$ ($29 \leq n \leq 43$), $C_nH_{2n-36}N_4V_1O_2$ ($31 \leq n \leq 46$), $C_nH_{2n-38}N_4V_1O_2$ ($31 \leq n \leq 44$)
3. $C_nH_{2n-30}N_4V_1O_3$ ($27 \leq n \leq 40$), $C_nH_{2n-32}N_4V_1O_3$ ($28 \leq n \leq 40$), $C_nH_{2n-34}N_4V_1O_3$ ($29 \leq n \leq 39$), $C_nH_{2n-36}N_4V_1O_3$ ($31 \leq n \leq 39$), $C_nH_{2n-38}N_4V_1O_3$ ($33 \leq n \leq 40$)
4. $C_nH_{2n-32}N_4V_1O_4$ ($28 \leq n \leq 36$), $C_nH_{2n-34}N_4V_1O_4$ ($30 \leq n \leq 35$), $C_nH_{2n-36}N_4V_1O_4$ ($n = 32$)

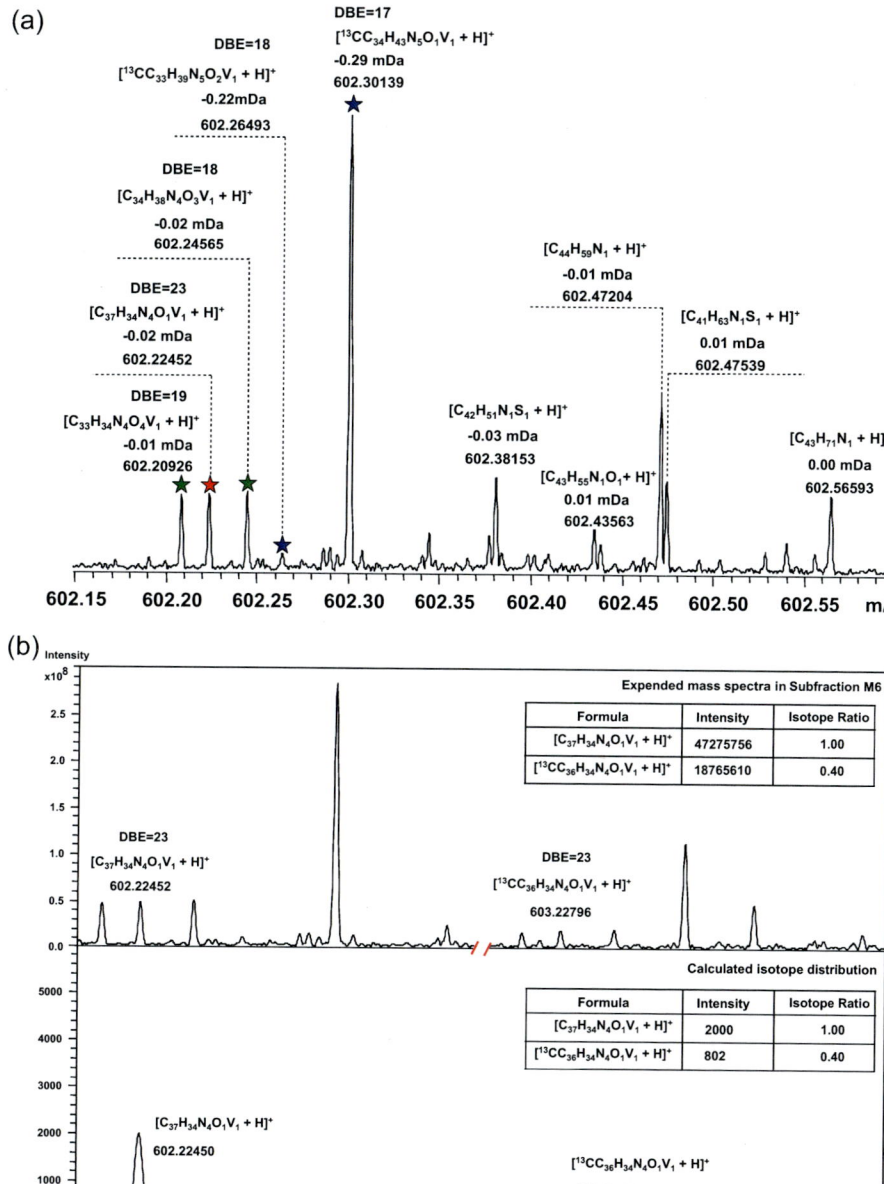

Fig. 4 (a) The expanded mass scale spectra of purification of vanadyl porphyrin in Venezuela heavy petroleum from the positive ion ESI FT-ICR MS at m/z 602; (b): the comparison chart between the real and calculated mass spectra for vanadyl porphyrin in Venezuela heavy petroleum from the positive ion ESI FT-ICR MS at m/z 602, 603

5. $C_nH_{2n-38}N_4V_1S_1O_1$ ($36 \leq n \leq 41$), $C_nH_{2n-40}N_4V_1S_1O_1$ ($35 \leq n \leq 51$), $C_nH_{2n-42}N_4V_1S_1O_1$ ($36 \leq n \leq 54$), $C_nH_{2n-44}N_4V_1S_1O_1$ ($41 \leq n \leq 55$), $C_nH_{2n-46}N_4V_1S_1O_1$ ($39 \leq n \leq 55$)
6. $C_nH_{2n-27}N_5V_1O_1$ ($29 \leq n \leq 40$), $C_nH_{2n-29}N_5V_1O_1$ ($34 \leq n \leq 42$), $C_nH_{2n-33}N_5V_1O_1$ ($31 \leq n \leq 38$), $C_nH_{2n-35}N_5V_1O_1$ ($32 \leq n \leq 41$)
7. $C_nH_{2n-27}N_5V_1O_2$ ($32 \leq n \leq 41$) and $C_nH_{2n-29}N_5V_1O_2$ ($33 \leq n \leq 42$)

Identification of these vanadium compounds was performed by assigning the spectral peaks to accurate mass values and isotopic masses and by observing their characteristic serial distribution at the large mass range. An exact mass match (within 0.5 mDa) is not sufficient to unambiguously identify the presence of vanadium compounds. The isotope ratio is critical to confirm the identification in addition to matching molecular mass. For example, Fig. 4 shows the expanded mass scale spectra at m/z 602 and the isotope ratio of $[C_{37}H_{34}N_4O_1V_1+H]^+$, the mass error is only 0.02 mDa, and both of the real and calculated isotope rations for $[^{13}CC_{36}H_{34}N_4O_1V_1+H]^+$ are 0.40 (the detailed analysis of other new vanadium compounds was described elsewhere [20, 88]). Therefore, good agreements were found not only in the accuracy mass but also in the isotope ratios. They were the powerful and significant evidences for these new vanadium compounds.

To identify the structures of these new vanadium compounds, the isolation experiments and collision-induced dissociation experiments were conducted using FT-ICR MS by Zhao et al. [88]. They speculated the possible structures of these new types of vanadium compounds, which are shown in Fig. 5. For Fig. 5a, they are the vanadyl porphyrins containing more fused aromatic rings and the functional groups of thiophene. Figure 5b–d shows the possible structures of three newly identified vanadium compounds based on the Treibs theoretical model on the formation of porphyrins from the evolution of chlorophyll and hemes. These molecular structures had varying DBE values and oxygen atoms, with functional groups of carbonyl and/or carboxyl at the peripheral of the porphyrin structure. Identification of $C_nH_mN_4V_1O_2$, $C_nH_mN_4V_1O_3$, and $C_nH_mN_4V_1O_4$ class species in crude oil is significant to the petroleum geochemistry and supports the hypothesis that porphyrins were derived from chlorophylls and hemes. McKenna et al. [89] also detected $C_nH_mN_4V_1O_2$ and $C_nH_mN_4V_1O_3$ by APPI FT-ICR MS in the petroleum. Figure 5e shows the new vanadium compounds with sulfur atoms; they may be generated from organic sulfur in the source kerogens or could be added by thermochemical sulfate reduction (TSR)/bacterial sulfate reduction (BSR) during the process of petroleum generation [90–93], which would convert side chains into condensed aromatic rings. Figure 5f, g shows the possible structures of the new vanadium compounds with five nitrogen atoms and one or two oxygen atoms. Hodgson [94] gave preliminary evidence for protein fragments associated with porphyrins, based on which the structures are reasonable. These new vanadium compounds contained the function group of amine and/or ether connected with the ETIO and DPEP porphyrin rings. These new compounds contain N, S, and O atoms which would enhance aggregation with asphaltene molecules in heavy petroleums.

Porphyrins in Heavy Petroleums: A Review

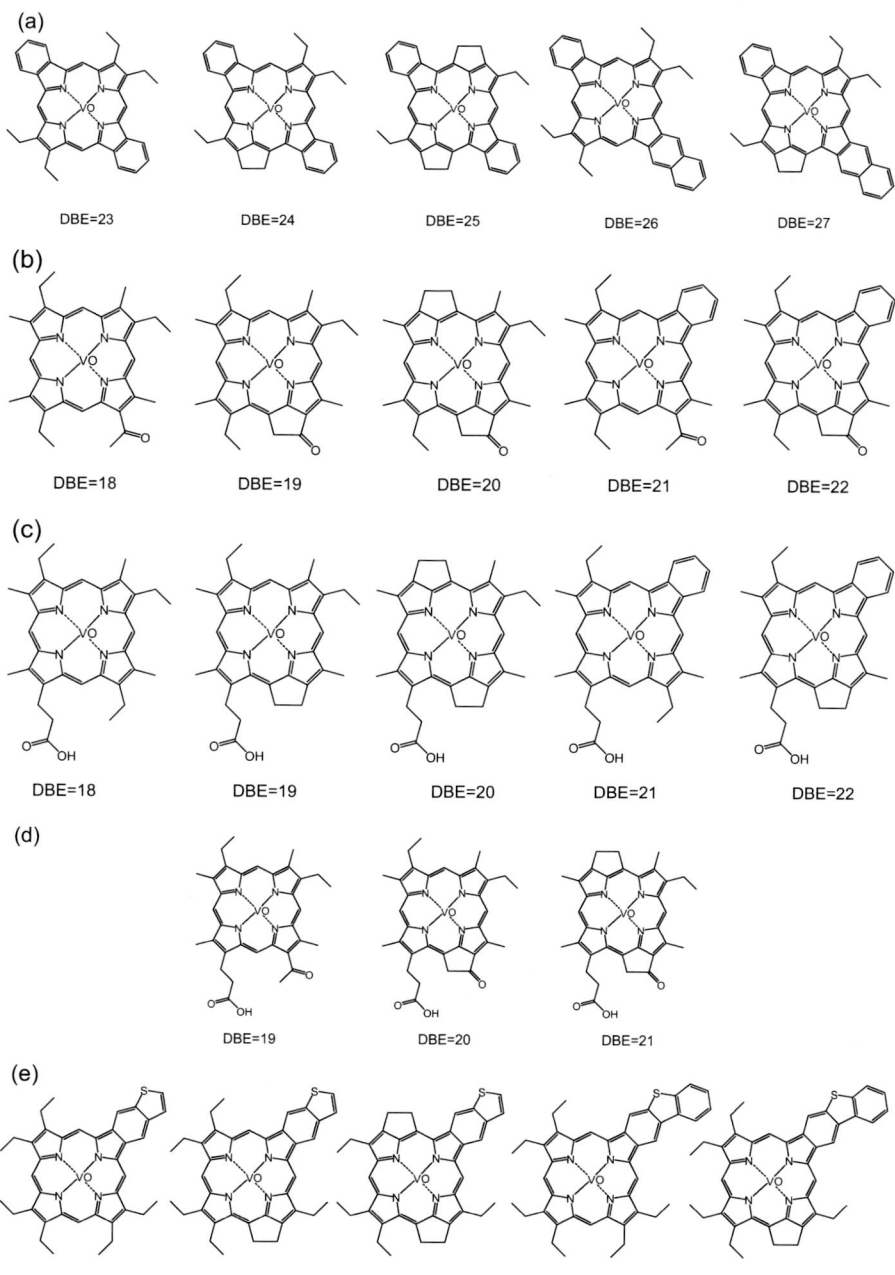

Fig. 5 (continued)

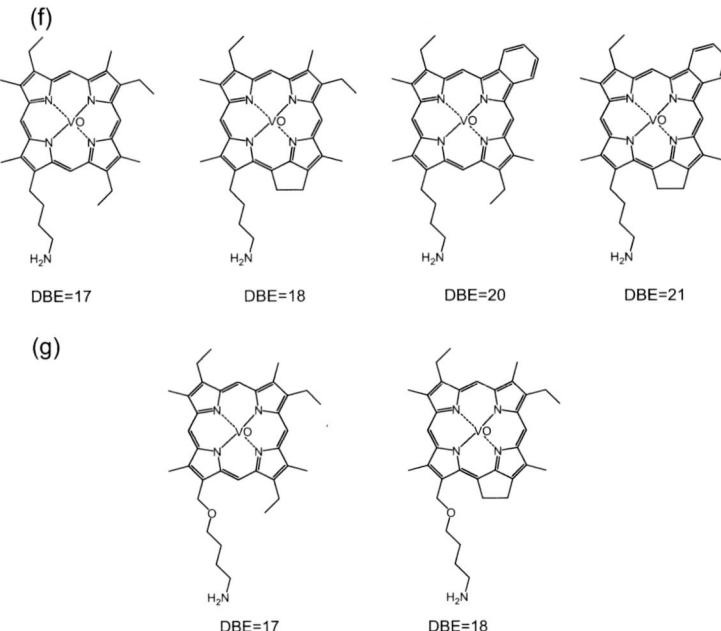

Fig. 5 The proposed structure of new class species of vanadium compounds in Venezuela crude oil detected in purified fractions by ESI FT-ICR MS. (**a**) for the vanadyl porphyrins containing more benzene rings linked the porphyrin ring, with different DBEs from 23 to 27; (**b–d**) for the vanadyl porphyrins with more oxygen atoms, with different DBEs from 18 to 22, 18 to 22, and 19 to 21, respectively, containing the function group of carbonyl and/or carboxyl; (**e**) for the vanadyl porphyrins with one sulfur, with different DBEs from 22 to 26, containing the functional group of thiophene; (**f**) for the vanadyl porphyrins with five nitrogen atoms and one oxygen atom, with different DBEs, 17, 18, 20, and 21, respectively, containing the function group of amino; (**g**) for the vanadyl porphyrins with five nitrogen atoms and two oxygen atoms, with different DBEs, 17 and 18, containing the function groups of amino and ether

To verify these possible structures, the molecular-level structural optimization has been investigated using the density functional theory (DFT) quantum chemical method, and calculations at the B3LYP and B3LYP/LanL2DZ/6-31 G++ level of theory using the Gaussian software were performed. The calculation results showed that these possible structures of new vanadium compounds could exist as stable entities.

These findings are significant for the understanding of the existing form of vanadium species in nature and are helpful for enhancing the amount of information on paleoenvironments and improving the level of applied basic theory for the processing technologies of heavy petroleums. Therefore, FT-ICR MS would be the most effective way to identify these vanadium compounds in a complex matrix, such as heavy petroleum.

4.3 Structure Characterization by X-Ray Absorption Spectroscopy

With the detection and identification of various series of porphyrins, UV–vis has been widely used in their quantification in petroleum samples. However, when the extinction coefficients of porphyrins are comparable to those of model compounds, generally VO-OEP, the quantification results measured by UV–vis were too small to account for the total vanadium or nickel content in petroleums [95, 96]. Yen et al. [97] found there were no characteristic absorption bands for the extraction of porphyrins; hence, they defined the "non-porphyrin" and pointed out that the difference between porphyrin and non-porphyrin is actually based on the properties of isolated fractions, rather than on structural types. Non-porphyrins will contain atypical porphyrins (such as aryl-porphyrin, reduced porphyrin, or decomposed porphyrin) or pseudo aromatic tetra dentate systems [98]. They also proposed the possible structures of non-porphyrins, such as containing the $N_3O_1(V_1O_1)$, $N_1O_3(V_1O_1)$, $N_2O_2(V_1O_1)$, $S_2O_2(V_1O_1)$, or $N_2S_2(V_1O_1)$.

X-ray absorption spectroscopy (XAFS) is a powerful method for characterizing the form of vanadium and nickel in petroleum; many authors used the extended X-ray absorption fine structure (EXAFS) spectroscopy and X-ray absorption near-edge structure (XANES) spectroscopy to obtain the bonding structure information around the vanadium atom and nickel atom in the petroleums.

Loos et al. [99] indicted that slight variations in the distances of V–N bond can be induced by different peripheral substituents; therefore, the EXAFS for the petroleum sample will include vanadyl porphyrins with different peripheral substitutions; the EXAFS will be an average spectrum. However, the XAFS spectra obtained by Goulon et al. [100] on different petroleum samples were strikingly similar to the pure VO-OEP; the vanadium is of the oxovanadyl type coordinated to four nitrogen atoms and in the porphyrin macrocycle. In spite of this, Poncet et al. [101] synthesized VO–ETIO compound with the four nitrogen atoms replaced by sulfur atoms; the XAFS spectrum of which was quite different from the VO-OEP spectrum. However, the XAFS spectra of vanadium compounds in petroleums were similar to traditional VO-OEP, unlikely to be coordinated to four sulfur atoms.

Miller et al. [102] separated the Maya n-heptane asphaltenes into "noncolloidal" asphaltenes and "colloidal" asphaltenes. The results of UV–vis showed that noncolloidal asphaltenes were present as metalloporphyrins and the colloidal asphaltenes were present as non-porphyrin, which showed no characteristic absorption bands of metalloporphyrins. However, both of their EXAFS spectra were similar to the spectrum of VO-OEP [103].

Because of the lower concentration of nickel in petroleums, there are not as many reports about the local structure of Ni non-porphyrins. Miller et al. [103] and Xu et al. [104] determined the local structure of Ni non-porphyrins in Maya residuum and Liaohe vacuum residue of China, respectively. The results indicated that the local coordination in Ni non-porphyrins was similar to that in Ni porphyrins.

In summary, although the petroleum samples in asphaltenes have shown that there were large amounts of "non-porphyrins," the vanadium is of the oxovanadyl type coordinated to four nitrogen atoms; as in the porphyrin macrocycle, the nickel is also of the type as Ni–N$_4$ in the porphyrin.

4.4 Other Methods

Actually, ultraviolet–visible (UV–vis)spectroscopy and mass spectrometry (MS) are indispensable and invaluable for the characterization of petroporphyrins. In addition, electron paramagnetic resonance spectroscopy (ESR) and chromatography with element detection such as size exclusion chromatography (SEC), GPC, or HPLC are also used for the characterization of petroporphyrins.

ESR could determine the resonant absorption of microwave radiation in the presence of a static magnetic field. Many authors have used EPR spectra to determine the bonding structure of the vanadium compounds in the heavy petroleums. Saraceno et al. [105] used VO–ETIO complex as a standard, and nominal EPR vanadium determinations were obtained on a series of distillated, residues, and full crudes having a total vanadium content in the range of 0.1 to 200 ppm, and the results were compared to values obtained by chemical analyses. They found a peak at the same characteristic location for vanadium. In addition, they obtained good agreement for the height of the peak with the standard; the linearity of the peak height vs. concentration plots and the similarity of the curve to those for VO–ETIO were found between the EPR determinations and the chemical results. It is indicated that the vanadium compounds in the petroleums were in the +4 valence state, such as in VO^{2+}. However, Yen et al. [97], Dickson et al. [106, 107], and Reynolds et al. [108, 109] concluded that a large portion of the vanadium compounds in petroleums were present in non-porphyrin coordination rather than the main structure of 4N(VO) based on EPR spin Hamiltonian parameters, although they were unable to define the non-porphyrin coordination type. It is notable that extremely high precision is needed when spin Hamiltonian parameters are used to determine the coordination structures around vanadium ion. Most of these measurements mentioned above have been made at 9 GHz using commercial EPR spectrometers. Malhotra et al. [110] used the high precision 34 GHz EPR to study the coordination of vanadyl complexes in various asphaltenes. It was concluded that non-porphyrin vanadyl spectra were not observed in Boscan asphaltene, and analysis of variance of the spin Hamiltonian parameters revealed that their ability to characterize the coordination of vanadyl complexes in bitumens and crude oil is limited. Therefore, EPR spectroscopy is not a suitable tool for identifying the ligand structure of vanadium in petroleum samples, such as vacuum residues and asphaltene.

As for the non-porphyrins, despite numerous studies directed toward their separation and identification, to date not a single molecular structure has been unambiguously confirmed. However, coupling techniques have proven highly effective with regard to the molecular characterization of trace metal compounds in complex petroleums, such as high-performance liquid chromatography combination with graphite furnace atomic absorption (HPLC-GFAA) [111], size-exclusion chromatography (SEC-HPLC) [112], and reversed-phase chromatography (RP-HPLC) [112, 113] in combination with inductively coupled or direct current plasma atomic emission spectroscopy (ICP or DCP).

When using SEC, due to the difference in retention times of different compounds within the column, the small molecules have longer retention times; therefore, the larger molecules are eluted from the column sooner. When using RP-HPLC, the highly polar molecules are also eluted sooner. Based on this premise, the petroleum samples were divided into several fractions based on molecular size and/or polarity, and they used chromatography combined with element detection to determine the molecule size, polar, and content of non-porphyrins in petroleums. Fish et al. [112] used RP-HPLC combination with GFAA to characterize the vanadyl and nickel non-porphyrin-rich fraction. On the basis of rapid-scan UV–vis data indicating a lack of Soret absorbance associated with separated vanadyl- and nickel-containing fractions, they have categorized highly polar nickel compounds as non-porphyrins in Boscan, Cerro Negro, Wilmington, and Prudhoe Bay heavy crude petroleums and confirmed that the lower polarity vanadyl non-porphyrin compounds were present in Wilmington and Prudhoe Bay heavy crude petroleums.

Caumette et al. [114] developed the coupling of SEC and normal phase (NP) HPLC using entirely organic mobile phases (tetrahydrofuran, xylene) with inductively coupled plasma mass spectrometry (ICP-MS) and investigated the molecular distribution of vanadium and nickel in crude oils. The metal species were fractionated by SEC using three columns in series with the increasing porosity (100, 1,000, and 100,000 Å) covering the molecular mass range (in eq. polystyrene) between 300 and 2×10^6 Da. NP-HPLC allowed the separation of the porphyrin-type fraction as well as separation of the remaining species into three distinct fractions. Because the metal species in the SEC fractions were found to be sufficiently stable to be collected, they also developed a bidimensional chromatography SEC-NP-HPLC–ICP- MS for the probing of the metal distribution in crude oils in terms of molecular weight and polarity. Their results indicated that most of vanadium and nickel were present in several families of non-porphyrins with different polarities and pointed out that vanadium complexes are less polar than nickel complexes, more vanadium complexes are present in low MW molecules than nickel complexes, and the presence of nickel is thus privileged in heavy and polar molecules.

The tendency for asphaltenes to associate in solution changes the apparent molecular size significantly. In addition, asphaltene fractions have a tendency to be adsorbed on the column (chemical interaction), and the elution order of polar fractions is significantly altered by the choice of solvent. Therefore, it should be serious to draw conclusions using SEC-HPLC to determine the molecular size and polarity of non-porphyrin. However, bidimensional chromatography offers a potentially purification method prior to the FT-ICR MS species identification. Till now, the application of hyphenated technique, such as HPLC-ICP MS or HPLC-FT-ICR

MS, would be the most effective way to identify vanadium and nickel compounds in a complex matrix, such as heavy petroleums.

5 Porphyrins as a Maturity Parameter for Petroleum Thermal Evolution

Porphyrins occur in petroleums and ancient sediments as vanadyl and nickel complexes. As mentioned, etioporphyrins (ETIO) and deoxophylloerythroetio porphyrins (DPEP) are the two most common petroporphyrins found in petroleums. Initially they were thought to be produced solely from chlorophyll a and heme, although the mode of entry of vanadium and nickel into the porphyrin structure had never been defined clearly [115, 116].

Biomarkers have been used both as oil–oil or oil–source rock correlation parameters and also as indicators of the thermal maturity of sediments and petroleums. Mackenzie et al. [117] observed the Toarcian shale of the Paris Basin followed by several reactions such as isomerization, aromatization, and cracking with the increasing of sediment maturity. During the research, steranes, aromatized steranes, triterpanes, isoprenoids, and porphyrins were studied. With the development and advances in HPLC and MS, porphyrins have been used to estimate the maturity of source rocks and petroleums. The ratio of DPEP/(ETIO+DPEP) has been used as a maturity indicator by several researchers [3, 40, 117–123]; it generally ranges from 0.8 in the least mature oils to 0 in the most mature oils. The ration decreases with increasing maturity. Didyk et al. [43] indicated that thermally induced conversion of DPEP to ETIO may be widespread in the geological environment and could account for the abundance of ETIO homologues in petroleums. However, Barwise et al. [124] proposed that the change in the ratio with maturity was due to the differences in the rate of decomposition; later, Barwise et al. [44] and Sundararaman et al. [45] suggested that kerogen contained amount of ETIO porphyrins and these are released during thermal cracking of kerogen; the change in the ratio with maturity was due to the dilution of preexisting DPEP porphyrin by ETIO porphyrins released from kerogen.

In addition, the porphyrin maturity parameter, PMP $= C_{28}E/(C_{28}E + C_{32}D)$, has been also used for maturity estimation of petroleums and source rocks [45], which was derived from the development of HPLC, where $C_{28}E$ is the ETIO porphyrin with 28 carbon atoms and $C_{32}D$ is the DPEP porphyrin with 32 carbon atoms. The PMP may be a more reliable maturity indicator for marine organic matter than some conventional methods. Sundararaman [119] investigated the cause for the change in PMP with maturity using isothermal and nonisothermal pyrolysis techniques; the results showed that it is principally due to the differences in the rates of decomposition of DPEP and ETIO porphyrins, and the release of bound ETIO porphyrins from the kerogen contributed to the change in PMP at the early stages of maturation.

Therefore, better understanding of the existing form of porphyrins species in nature is helpful for enhancing the amount of information on paleoenvironments, and the decrease in the ratio of DPEP/ETIO with increasing maturation could be useful to determine the extent of maturation of genetically related geological materials.

6 Demetallization Technologies for Petroleum Upgrading

Vanadium and nickel compounds in heavy crude oil existed as organic forms and are difficult to be remove in the electric desalting processes, and these compounds have significant detrimental impacts during upgrading processes, such as the catalytic cracking and catalytic hydrogenation process, particularly in the catalyst poisoning. In the catalytic cracking unit, these metal contaminants change the activity–selectivity rating of the unit's catalyst inventory. There are several technologies for demetallization and upgrading which were developed since the beginning of the petroleum industry, e.g., physical methods, chemical methods, coking, and catalytic hydroprocessing.

The solvent deasphalting process is a typical physical method to remove the metals in heavy petroleums. Deasphalting with liquid hydrocarbon or gasses such as propane, butane, or isobutene is very effective and used by refineries for removal of metals from vacuum residues. However, the traditional process also removes a large amount of convertible material along with the metal-containing species [125]; in addition, large amounts of solvent and the energy consumption required to recover the solvent are needed. In downstream upgrading, most refineries are ill equipped to process heavy crude oils, because the concentrations of contaminants in this feedstock are higher and the capability for by-product disposal may be inadequate. The nickel, vanadium, and nitrogen contents in heavy crude oils are much higher. These components are among the most intractable in catalytic upgrading processes and will have a significant adverse impact, especially on catalytic processes. In order to improve the solvent deasphalting process, Zhao et al. [126, 127] have developed supercritical fluid extraction and fractionation (SFEF) to separate a variety of petroleum residua and other heavy petroleums into narrow-cut fractions with total yields up to 75–90%. By using SFEF to cut heavy oils, the products are much cleaner and less viscous than the feedstock, while the end cut with most of the contaminants, such as sulfur, Ni, V, and asphaltene could be discarded. A demonstration plant in PetroChina Liaohe petrochemical company has shown that selective extraction of asphaltenes technology can be built and operated with low technical and commercial risk as selective extraction of asphaltene technology [127]. Fractionation of residues, or pitches, by SFEF is a useful diagnostic tool for validation of bitumen upgrading processes. Because the solvent polarity in SFEF is readily controllable, this technique also offers the most flexible approach to bitumen pitch separation. In addition, the SFEF end-cut fraction, which was not useful products with most of the contaminants, could rather be considered as a potential feed for gasification or as a source of value-added products.

As for the chemical methods, the basic chemical concept of demetallization is to selectively remove the metal from the organic moiety with minimal conversion of the remaining petroleum. Treatment of petroleum fractions with sulfuric acid has been used commercially for many years at the very beginning of petroleum industry. However, acids have some disadvantages, including extensive side reactions and product contamination.

Actually, the technology of coking is also widely used to process heavy oils. The process could quantitatively capture the metals in the coke [128]. In addition, the process of residue fluid catalytic cracking (RFCC) could also capture the vanadium and nickel in the coke, which deposits on the catalyst pellets. However, the catalyst structure would be destroyed during the burned regeneration of catalyst [129, 130].

Catalytic hydroprocessing is a hydrogenation process; it selectively removes the metals from liquid petroleum fractions by the process of hydrodemetallization (HDM), where the porphyrins lose the metal atoms by hydrogenation reactions. It is generally accepted that the HDM mechanism of porphyrins consists of two sequential steps, the first being the reversible hydrogenation of peripheral double bonds resulting in the formation of metallochlorins and the second being the irreversible hydrogenolysis step on the methine position which leads to the fragmentation of the big ring and the removal of central metals [131–134]. The substitution pattern on the periphery of porphyrins was found to affect their HDM reactivity: the hydrogenation of peripheral bonds is strongly influenced by the substituents on the b-pyrrolic position, and the hydrogenolysis activity is related to the substituents on the methine position [131–133]. The analyses indicate that porphyrins are series of alkyl-substituted porphyrins. For such porphyrins, the electron-donating ability of the alkyl groups renders the b-pyrrolic position more basic, and at the same time, the alkyl groups hinder porphyrin molecule access to hydrogenation centers of catalyst. Hence, as the content of porphyrins with high carbon numbers increases, their hydrogenation reactivity decreases. In contrast, if there is not any substituent on the methine position, it can be easily attacked and the hydrogenolysis reactivity will thus be high. As a result, after the preliminary hydrogenation, porphyrins of ETIO type become more easily hydrogenolysized than that of DPEP type. In view of the diversity of metal compounds and their reactivity, a series of HDM catalysts with different activities and the corresponding catalyst grading system are used for vacuum residue desulfurization (VRDS) unit, thus not only removing most metals from petroleum but also leading to a uniform distribution of metal deposit within the reactor [135]. The products of HDM can accumulate in the catalyst pores, causing the formation of deposits, which accumulates in the catalyst and eventually renders it inactive[8, 131, 136, 137]. Therefore, high metal content feeds require high addition rates of the catalyst in ebullated bed reactors [138].

In addition, oxidative demetallization [139], biological demetallization [140], electrolytic demetallization [141], ultrasonic irradiation and adsorption [142], photochemical reaction and liquid extraction [143], and absorption by Mo complexes

[144] have been also explored. However, only catalytic demetallization was a selective removal technique.

7 Summary and Future Prospects

As discussed above, although numerous new series of vanadyl porphyrins have been identified recently, there are still more vanadium and nickel compounds which have not been detected. That is because of the low concentration of porphyrins and the complexity of asphaltenes in heavy petroleums. Therefore, effective separation is the prerequisite; and the research of quantitative methods is still one of the most important directions.

In addition, ultrahigh mass resolution is also needed to indentify these vanadium and nickel compounds. FT-ICR MS is still the significant method of qualitative analysis for vanadium and nickel compounds. Better qualitative and quantitative understanding of porphyrins in oil residues will be helpful for enhancing the amount of information on paleoenvironments and improving the level of applied basic theory for the processing technologies of heavy petroleums.

Acknowledgement This work was supported by the National Natural Science Foundation of China (NSFC, 21376262, 21236009) and National Basic Research Program of China (2010CB226901). The authors declare no competing financial interests.

References

1. Meyer RF, Attanasi ED, Freeman PA (2007) Heavy oil and natural bitumen resources in geological basins of the world
2. Xu C, Bell L (2013) Worldwide reserves, oil production post modest rise. Oil Gas J 111(12):30–31
3. Branthaver JF (1987) Metal complexes in fossil fuels: geochemistry, characterization, and processing. In: Filby RH, Branthaver JF (ed.) ACS Symposium Series 344, Washington, D.C. American Chemical Society, pp 188–204
4. Treibs A (1934) Chlorophyll and haemin derivatives in bituminous rocks, petroleum, mineral waxes and asphalts. Justus Liebig's Annalen Chem 510:42–62
5. Treibs A (1936) Chlorophyll and hemin derivatives in organic materials. Angew Chem 49(38):682–686
6. Stoyanov SR, Yin C-X, Gray MR, Stryker JM, Gusarov S, Kovalenko A (2010) Computational and experimental study of the structure, binding preferences, and spectroscopy of nickel (II) and vanadyl porphyrins in petroleum. J Phys Chem B 114(6):2180–2188
7. Mckenna AM, Purcell JM, Rodgers RP, Marshall AG (2009) Identification of vanadyl porphyrins in a heavy crude oil and raw asphaltene by atmospheric pressure photoionization fourier transform ion cyclotron resonance (FT-ICR) mass spectrometry. Energy Fuel 23(4):2122–2128
8. Agrawal R, Wei J (1984) Hydrodemetalation of nickel and vanadium porphyrins. 1. Intrinsic kinetics. Ind Eng Chem Process Design Dev 23(3):505–514

9. Lewan M, Maynard J (1982) Factors controlling enrichment of vanadium and nickel in the bitumen of organic sedimentary rocks. Geochim Cosmochim Acta 46(12):2547–2560
10. Lewan MD (1984) Factors controlling the proportionality of vanadium to nickel in crude oils. Geochim Cosmochim Acta 48(11):2231–2238
11. Barwise A (1990) Role of nickel and vanadium in petroleum classification. Energy Fuel 4(6):647–652
12. Xu H, Que G, Yu D, Lu JR (2005) Characterization of petroporphyrins using ultraviolet-visible spectroscopy and laser desorption ionization time-of-flight mass spectrometry. Energy Fuel 19(2):517–524
13. Barwise A, Whitehead E (1980) Separation and structure of petroporphyrins. Phys Chem Earth 12:181–192
14. Reynolds JG (1985) Characterization of heavy residua by application of a modified D2007 separation and electron paramagnetic resonance. Liquid Fuels Technol 3(1):73–105
15. Zhang L, Xu Z, Shi Q, Sun X, Zhang N, Zhang Y, Chung KH, Xu C, Zhao S (2012) Molecular characterization of polar heteroatom species in venezuela orinoco petroleum vacuum residue and its supercritical fluid extraction subfractions. Energy Fuel 26(9):5795–5803
16. Pearson CD, Green JB (1989) Comparison of processing characteristics of Mayan and Wilmington heavy residues: 2. characterization of vanadium and nickel complexes in acid-base-neutral fractions. Fuel 68(4):465–474
17. Baker E, Palmer S (1978) Geochemistry of porphyrins. Porphyrins 1:485–551
18. Hajibrahim S, Tibbetts P, Watts C, Maxwell J, Eglinton G, Colin H, Guiochon G (1978) Analysis of carotenoid and porphyrin pigments of geochemical interest by high-performance liquid chromatography. Anal Chem 50(4):549–553
19. Chen P, Xing Z, Liu M, Liao Z, Huang D (1999) Isolation of nine petroporphyrin biomarkers by reversed-phase high-performance liquid chromatography with coupled columns. J Chromatogr A 839(1):239–245
20. Zhao X, Liu Y, Xu C, Yan Y, Zhang Y, Zhang Q, Zhao S, Chung K, Gray MR, Shi Q (2013) Separation and characterization of vanadyl porphyrins in Venezuela Orinoco heavy crude oil. Energy Fuel 27(6):2874–2882
21. Martin J, Quirke E, Shaw GJ, Soper PD, Maxwell JR (1980) Petroporphyrins—II: the presence of porphyrins with extended alkyl substituents. Tetrahedron 36(22):3261–3267
22. Dunning H, Rabon NA (1956) Porphyrin-metal complexes in petroleum stocks. Ind Eng Chem 48(5):951–955
23. Baker EW (1966) Mass spectrometric characterization of petroporphyrins1. J Am Chem Soc 88(10):2311–2315
24. Quirke JME, Eglinton G, Maxwell JR (1979) Petroporphyrins. 1. Preliminary characterization of the porphyrins of gilsonite. J Am Chem Soc 101(26):7693–7697
25. Kowanko N, Branthaver JF, Sugihara JM (1978) Direct liquid-phase fluorination of petroleum. Fuel 57(12):769–775
26. Branthaver J, Trudell L, Heppner R (1982) Nickel porphyrins in the bedding planes of a Colorado Lean Oil Shale. Org Geochem 4(1):1–7
27. Branthaver J, Storm C, Baker E (1983) An investigation of the structure of abelsonites from the Uinta Basin of Utah. Org Geochem 4(3):121–134
28. Ali MF, Perzanowski H, Bukhari A, Al-Haji AA (1993) Nickel and vanadyl porphyrins in Saudi Arabian crude oils. Energy Fuel 7(2):179–184
29. Ekstrom A, Fookes C, Hambley T, Loeh H, Miller S, Taylor J (1983) Determination of the crystal structure of a petroporphyrin isolated from oil shale. Nature 306:173–174
30. Baker EW, Louda JW (1984) Highly dealkylated copper and nickel etioporphyrins in marine sediments. Org Geochem 6:183–192
31. Erdman JG (1965) Process for removing metals from a mineral oil with an alkyl sulfonic acid. US Patent 3,190,829
32. Baker EW, Yen TF, Dickie JP, Rhodes RE, Clark LF (1967) Mass spectrometry of porphyrins. II. Characterization of petroporphyrins. J Am Chem Soc 89(14):3631–3639

33. Popl M, Dolanský V, Šebor G, Stejskal M (1978) Hydrocarbons and porphyrins in rock extracts. Fuel 57(9):565–570
34. Van Berkel GJ, Quirke JME, Filby RH (1989) The Henryville bed of the new albany shale—I. Preliminary characterization of the nickel and vanadyl porphyrins in the bitumen. Org Geochem 14(2):119–128
35. Chakraborty S, Bhatia V (1981) Isolation and characterization of metalloporphyrins from darius crude. Indian J Technol 19(3):92–99
36. Eglinton G, Hajibrahim SK, Maxwell JR, Martin J, Quirke E (1980) Petroporphyrins: structural elucidation and the application of HPLC fingerprinting to geochemical problems. Phys Chem Earth 12:193–203
37. Kashiyama Y, Kitazato H, Ohkouchi N (2007) An improved method for isolation and purification of sedimentary porphyrins by high-performance liquid chromatography for compound-specific isotopic analysis. J Chromatogr A 1138(1):73–83
38. Hajibrahim SK (1981) Development of high pressure liquid chromatography (HPLC) for fractionation and fingerprinting of petroporphyrin mixtures. J Liquid Chromatogr 4(5):749–764
39. Hohn M, Hajibrahim S, Eglinton G (1982) High-pressure liquid chromatography of petroporhyrins: evaluation as a geochemical fingerprinting method by principal components analysis. Chem Geol 37(3):229–237
40. Sundararaman P, Raedeke LD (1993) Vanadyl porphyrins in exploration: maturity indicators for source rocks and oils. Appl Geochem 8(3):245–254
41. Quirke J, Eglinton G, Palmer S, Baker E (1982) High-performance liquid chromatographic and mass spectrometric analyses of porphyrins from deep-sea sediments. Chem Geol 35(1):69–85
42. Peng P, Eglinton G, Fu J, Sheng G (1992) Biological markers in chinese ancient sediments. 1. Geoporphyrins. Energy Fuel 6(2):215–225
43. Didyk BM, Alturki YI, Pillinger CT, Eglinton G (1975) Petroporphyrins as indicators of geothermal maturation. Nature, 256:563–565
44. Barwise AJG (1987) Mechanisms involved in altering deoxophylloerythroetioporphyrin-etioporphyrin ratios in sediments and oils. In: Filby RH, Branthaver JF (ed.) ACS Symposium Series 344, Washington, D.C. American Chemical Society, pp 100–109
45. Sundararaman P, Biggs WR, Reynolds JG, Fetzer JC (1988) Vanadylporphyrins, indicators of kerogen breakdown and generation of petroleum. Geochim Cosmochim Acta 52(9):2337–2341
46. Doukkali A, Saoiabi A, Zrineh A, Hamad M, Ferhat M, Barbe J, Guilard R (2002) Separation and identification of petroporphyrins extracted from the oil shales of Tarfaya: geochemical study. Fuel 81(4):467–472
47. Xu H, Yu D, Que G (2005) Characterization of petroporphyrins in gudao residue by ultraviolet–visible spectrophotometry and laser desorption ionization-time of flight mass spectrometry. Fuel 84(6):647–652
48. Glebovskaya E, Volkenshtein M (1948) Spectra of porphyrins in petroleums and bitumens. J Gen Chem (USSR) 18:1440
49. Skinner DA (1952) Chemical state of vanadium in Santa Maria Valley crude oil. Ind Eng Chem 44(5):1159–1165
50. Hood A, Carlson E, O'neal M (1960) Petroleum oil analysis. In: Clark GL (ed) Encyclopedia of spectroscopy. New York, pp 613
51. Mead W, Wilde A (1961) Mass spectrum of vanadyl etioporphyrin-I. Chem Ind 33:1315–1316
52. Freeman DH, Swahn ID, Hambright P (1990) Spectrophotometry and solubility properties of nickel and vanadyl porphyrin complexes. Energy Fuel 4(6):699–704
53. Freeman DH, Saint Martin DC, Boreham CJ (1993) Identification of metalloporphyrins by third-derivative Uv/Vis diode array spectroscopy. Energy Fuel 7(2):194–199

54. Foster NS, Day JW, Filby RH, Alford A, Rogers D (2002) The role of Na-montmorillonite in the evolution of copper, nickel, and vanadyl geoporphyrins during diagenesis. Org Geochem 33(8):907–919
55. Cantú R, Stencel JR, Czernuszewicz RS, Jaffé PR, Lash TD (2000) Surfactant-enhanced partitioning of nickel and vanadyl deoxophylloerythroetioporphyrins from crude oil into water and their analysis using surface-enhanced resonance raman spectroscopy. Environ Sci Technol 34(1):192–198
56. Freeman DH, O'haver TC (1990) Derivative spectrophotometry of petroporphyrins. Energy Fuel 4(6):688–694
57. Gallegos EJ, Sundararaman P (1985) Mass spectrometry of geoporphyrins. Mass Spectrom Rev 4(1):55–85
58. Howe WW (1961) Improved chromatographic analysis of petroleum porphyrin aggregates and quantitative measurement by integral absorption. Anal Chem 33(2):255–260
59. Thomas D, Blumer M (1964) Porphyrin pigments of a triassie sediment. Geochim Cosmochim Acta 28(7):1147–1154
60. Yen TF (1975) Chemical aspects of metals in native petroleum. In: Yen TF (ed.) The Role of Trace Metals in Petroleum. Ann Arbor Science Publishers, pp 1–30
61. Qi L, Liang R, Wang X (1981) Study of nickel porphyrins in some chinese crude oils. Acta Petrolei Sinica 2(4):108–116
62. Prowse W, Chicarelli M, Keely B, Kaur S, Maxwell J (1987) Characterisation of fossil porphyrins of the "Di-Dpep" type. Geochim Cosmochim Acta 51(10):2875–2877
63. Liao Z, Huang D, Shi J (1990) Discovery of special predominance of vanadyl porphyrin and high abundance of Di-Dpep in nonmarine strata. Scientia Sinica Ser B 33(5):631–640
64. Quirke J (1987) Techniques for isolation and characterization of the geoporphyrins and chlorines. In: Filby RH, Branthaver JF (ed.) ACS Symposium Series 344, Washington, D.C. American Chemical Society, pp 308–337
65. Shaw GJ, Quirke JME, Eglinton G (1978) Analysis of petroporphyrins by chemical ionisation mass spectrometry. J Chem Soc Perkin Trans 1(12):1655–1659
66. Grigsby R, Green J (1997) High-resolution mass spectrometric analysis of a vanadyl porphyrin fraction isolated from the >700°C Resid of Cerro Negro Heavy Petroleum. Energy Fuel 11(3):602–609
67. Premović P, Đorđević D, Pavlović M (2002) Vanadium of petroleum asphaltenes and source kerogens (La Luna Formation, Venezuela): isotopic study and origin. Fuel 81(15):2009–2016
68. Wright BW, Smith RD (1989) Supercritical fluid chromatography-mass spectrometry: a potentially useful technique for porphyrin analysis. Org Geochem 14(2):227–232
69. Eckardt CB, Dyas L, Yendle PW, Eglinton G (1988) Multimolecular data processing and display in organic geochemistry: the evaluation of petroporphyrin GC-MS data. Org Geochem 13(4):573–582
70. Mcfadden W, Bradford D, Eglinton G, Hajlbrahim S, Nicolaides N (1979) Application of combined liquid chromatography/mass spectrometry (LC/MS): analysis of petroporphyrins and meibomian gland waxes. J Chromatogr Sci 17(9):518–522
71. Johnson JV, Britton ED, Yost RA, Quirke JME, Cuesta LL (1986) Tandem mass spectrometry for characterization of high-carbon-number geoporphyrins. Anal Chem 58 (7):1325–1329
72. Beato BD, Yost RA, Van Berkel GJ, Filby RH, Quirke JME (1991) The Henryville bed of the new albany shale—III: tandem mass spectrometric analyses of geoporphyrins from the bitumen and kerogen. Org Geochem 17(1):93–105
73. Van Berkel GJ, Quinones MA, Quirke JME (1993) Geoporphyrin analysis using electrospray ionization-mass spectrometry. Energy Fuel 7(3):411–419
74. Fenn JB, Mann M, Meng CK, Wong SF, Whitehouse CM (1989) Electrospray ionization for mass spectrometry of large biomolecules. Science 246(4926):64–71
75. Fenn JB, Mann M, Meng CK, Wong SF, Whitehouse CM (1990) Electrospray ionization–principles and practice. Mass Spectrom Rev 9(1):37–70

76. Iribarne J, Thomson B (1976) On the evaporation of small ions from charged droplets. J Chem Phys 64(6):2287–2294
77. Thomson B, Iribarne J (1979) Field induced ion evaporation from liquid surfaces at atmospheric pressure. J Chem Phys 71(11):4451–4463
78. Sakairi M, Yergey AL, Siu KM, Le Blanc JY, Guevremont R, Berman SS (1991) Electrospray mass spectrometry: application of ion evaporation theory to amino acids. Anal Chem 63(14):1488–1490
79. Fenn JB, Rosell J, Meng CK (1997) In electrospray ionization, how much pull does an ion need to escape its droplet prison? J Am Soc Mass Spectrom 8(11):1147–1157
80. Cech NB, Enke CG (2001) Practical implications of some recent studies in electrospray ionization fundamentals. Mass Spectrom Rev 20(6):362–387
81. Marshall AG, Hendrickson CL (2008) High-resolution mass spectrometers. Annu Rev Anal Chem 1:579–599
82. Marshall AG, Rodgers RP (2004) Petroleomics: the next grand challenge for chemical analysis. Acc Chem Res 37(1):53–59
83. Rodgers RP, Schaub TM, Marshall AG (2005) Petroleomics: Ms returns to its roots. Anal Chem 77(1):20A–27A
84. Rodgers RP, Hendrickson CL, Emmett MR, Marshall AG, Greaney M, Qian K (2001) Molecular characterization of petroporphyrins in crude oil by electrospray ionization fourier transform ion cyclotron resonance mass spectrometry. Can J Chem 79(5-6):546–551
85. Purcell JM, Hendrickson CL, Rodgers RP, Marshall AG (2006) Atmospheric pressure photoionization fourier transform ion cyclotron resonance mass spectrometry for complex mixture analysis. Anal Chem 78(16):5906–5912
86. Qian K, Mennito AS, Edwards KE, Ferrughelli DT (2008) Observation of vanadyl porphyrins and sulfur–containing vanadyl porphyrins in a petroleum asphaltene by atmospheric pressure photoionization fourier transform ion cyclotron resonance mass spectrometry. Rapid Commun Mass Spectrom 22(14):2153–2160
87. Qian K, Edwards KE, Mennito AS, Walters CC, Kushnerick JD (2009) Enrichment, resolution, and identification of nickel porphyrins in petroleum asphaltene by cyclograph separation and atmospheric pressure photoionization fourier transform ion cyclotron resonance mass spectrometry. Anal Chem 82(1):413–419
88. Zhao X, Shi Q, Gray MR, Xu C (2014) New vanadium compounds in venezuela heavy crude oil detected by positive-ion electrospray ionization fourier transform ion cyclotron resonance mass spectrometry. Sci. Rep., 4:5373
89. Mckenna AM, Williams JT, Putman JC, Aeppli C, Reddy CM, Valentine DL, Lemkau KL, Kellermann MY, Savory JJ, Kaiser NK (2014) Unprecedented ultrahigh resolution FT-ICR mass spectrometry and parts-per-billion mass accuracy enable direct characterization of nickel and vanadyl porphyrins in petroleum from natural seeps. Energy Fuel 28(4):2454–2464
90. Jørgensen B (1977) Bacterial sulfate reduction within reduced microniches of oxidized marine sediments. Mar Biol 41(1):7–17
91. Orr WL (1986) Kerogen/asphaltene/sulfur relationships in sulfur-rich monterey oils. Org Geochem 10(1):499–516
92. Strausz OP, Mojelsky TW, Lown EM (1992) The molecular structure of asphaltene: an unfolding story. Fuel 71(12):1355–1363
93. Peters KE, Fowler MG (2002) Applications of petroleum geochemistry to exploration and reservoir management. Org Geochem 33(1):5–36
94. Hodgson G, Baker B (1969) Porphyrins in meteorites: metal complexes in orgueil, murray, cold bokkeveld, and mokoia carbonaceous chondrites. Geochim Cosmochim Acta 33(8):943–958
95. Sugihara JM, Bean RM (1962) Direct determination of metalloporphyrins in boscan crude oil. J Chem Eng Data 7(2):269–271

96. Senglet N, Williams C, Faure D, Des Courieres T, Guilard R (1990) Microheterogeneity study of heavy crude petroleum by UV–Visible spectroscopy and small angle X-ray scattering. Fuel 69(1):72–77
97. Yen TF, Boucher LJ, Dickie JP, Tynan EC, Vaughan GB (1969) Vanadium complexes and porphyrins in asphaltenes. J Inst Petroleum 55(542):87–93
98. Yen T (1978) The nature of vanadium complexes in the refining of heavy oil. Energy Sources 3(3-4):339–351
99. Loos M, Ascone I, Friant P, Ruiz-Lopez M, Goulon J, Barbe J, Senglet N, Guilard R, Faure D, Des Courieres T (1990) Vanadyl porphyrins: evidence for self-association and for specific interactions with hydroprocessing catalysts shown from XAFS and ESR studies. Catal Today 7(4):497–513
100. Goulon J, Retournard A, Friant P, Goulon-Ginet C, Berthe C, Muller J-F, Poncet J-L, Guilard, R, Escalier J-C, Neff B (1984) Structural characterization by X-Ray absorption spectroscopy (exafs/xanes) of the vanadium chemical environment in boscan asphaltenes. J Chem Soc Dalton Trans (6):1095–1103
101. Poncet JL, Guilard R, Friant P, Goulonginet C, Goulon J (1984) Vanadium (IV) porphyrins-synthesis and classification of thiovanadyl and selenovanadyl porphyrins-EXAFS and RPE spectroscopic studies. New J Chem 8(10):583–590
102. Miller J, Fisher R, Thiyagarajan P, Winans R, Hunt J (1998) Subfractionation and characterization of Mayan asphaltene. Energy Fuel 12(6):1290–1298
103. Miller J, Fisher R, Van Der Eerden A, Koningsberger D (1999) Structural determination by XAFS spectroscopy of non-porphyrin nickel and vanadium in Maya residuum, hydrocracked residuum, and toluene-insoluble solid. Energy Fuel 13(3):719–727
104. Xu H, Que G-H, Wang J-Q, Yu D-Y, Zhang J, Xie Y-N, Hu T-D (2003) Structural characterization by XAFS spectroscopy of non-porphyrin nickel in liaohe vacuum residue. Acta Chim Sinica 61(3):450–453
105. Saraceno A, Fanale D, Coggeshall N (1961) An electron paramagnetic resonance investigation of vanadium in petroleum oils. Anal Chem 33(4):500–505
106. Dickson F, Kunesh C, Mcginnis E, Petrakis L (1972) Use of electron spin resonance to characterize the vanadium (IV)-sulfur species in petroleum. Anal Chem 44(6):978–981
107. Dickson FE, Petrakis L (1974) Application of electron spin resonance and electronic spectroscopy to the characterization of vanadium species in petroleum fractions. Anal Chem 46(8):1129–1130
108. Reynolds JG, Biggs WR, Fetzer JC (1985) Characterization of vanadium compounds in selected crudes II. Electron paramagnetic resonance studies of the first coordination spheres in porphyrin and non-porphyrin fractions. Liquid Fuels Technol 3(4):423–448
109. Reynolds JG, Gallegos EJ, Fish RH, Komlenic JJ (1987) Characterization of the binding sites of vanadium compounds in heavy crude petroleum extracts by electron paramagnetic resonance spectroscopy. Energy Fuel 1(1):36–44
110. Malhotra VM, Buckmaster HA (1985) 34 Ghz EPR study of vanadyl complexes in various asphaltenes: statistical correlative model of the coordinating ligands. Fuel 64(3):335–341
111. Fish RH, Komlenic JJ (1984) Molecular characterization and profile identifications of vanadyl compounds in heavy crude petroleums by liquid chromatography/graphite furnace atomic absorption spectrometry. Anal Chem 56(3):510–517
112. Fish RH, Komlenic JJ, Wines BK (1984) Characterization and comparison of vanadyl and nickel compounds in heavy crude petroleums and asphaltenes by reverse-phase and size-exclusion liquid chromatography/graphite furnace atomic absorption spectrometry. Anal Chem 56(13):2452–2460
113. Fish R, Izquierdo A, Komlenic J, Reynolds J, Gallegos E (1986) Molecular characterization of nickel and vanadium non-porphyrin compounds found in heavy crude petroleums and bitumens. Am Chem Soc Div Pet Chem Prepr (United States) 31:CONF-860425

114. Caumette G, Lienemann C-P, Merdrignac I, Bouyssiere B, Lobinski R (2010) Fractionation and speciation of nickel and vanadium in crude oils by size exclusion chromatography-ICP MS and normal phase HPLC-ICP MS. J Anal Atomic Spectrometry 25(7):1123–1129
115. Hodgson G, Peake E (1961) Metal chlorine complexes in recent sediments as initial precursors to petroleum porphyrin pigments. Nature, 191(4790):766–767
116. Hodgson G (1973) Geochemistry of porphyrins—reactions during diagenesis. Ann N Y Acad Sci 206(1):670–684
117. Mackenzie JQ Jr (1980) Maxwell, molecular parameters of maturation in the toarcian shales, Paris Basin, France, II. Evolution of metallo-porphyrins. In: Ag Douglas JM (ed) Advances in organic geochemistry. Pergamon, Oxford, pp 239–248
118. Barwise A (1983) Use of porphyrins as a maturity parameter for oils and sediments. Geol Soc Lond Spec Publ 12(1):309–315
119. Sundararaman P (1993) On the mechanism of change in DPEP/ETIO ratio with maturity. Geochim Cosmochim Acta 57(18):4517–4520
120. Sundararaman P, Dahl JE (1993) Depositional environment, thermal maturity and irradiation effects on porphyrin distribution: Alum Shale, Sweden. Org Geochem 20(3):333–337
121. Sundararaman P, Moldowan JM (1993) Comparison of maturity based on steroid and vanadyl porphyrin parameters: a new vanadyl porphyrin maturity parameter for higher Maturities. Geochim Cosmochim Acta 57(6):1379–1386
122. Lee AK, Gregg HR, Reynolds JG (1995) Metallopetroporphyrins as process indicators: mass spectral identification of NI (ETIO) and NI (DPEP) homologous series in Green River Shale Oil. Fuel Sci Tech Int 13(9):1153–1166
123. Lee AK, Murray AM, Reynolds JG (1995) Metallopetroporphyrins as process indicators: separation of petroporphyrins in Green River Oil Shale pyrolysis products. Fuel Sci Technol Int 13(8):1081–1097
124. Barwise A, Roberts I (1984) Diagenetic and catagenetic pathways for porphyrins in sediments. Org Geochem 6:167–176
125. Brons G, Yu JM (1995) Solvent deasphalting effects on whole cold lake bitumen. Energy Fuel 9(4):641–647
126. Zhao S, Xu Z, Xu C, Chung KH, Wang R (2005) Systematic characterization of petroleum residua based on SFEF. Fuel 84(6):635–645
127. Zhao S, Xu C, Sun X, Chung KH, Xiang Y (2010) China refinery tests asphaltenes extraction process. Oil Gas J 108(12):52–59
128. Siskin M, Kelemen S, Eppig C, Brown L, Afeworki M (2006) Asphaltene molecular structure and chemical influences on the morphology of coke produced in delayed coking. Energy Fuel 20(3):1227–1234
129. Speronello B, Reagan W (1984) Test measures FCC catalyst deactivation by Ni V. Oil Gas J 82(5):139–143
130. Tsiatouras VA, Evmiridis NP (2008) FCC catalysts: Cu (II)-exchanged USY-type. Stability, dealumination, and acid sites after thermal and hydrothermal treatment before and after vanadium impregnation. Ind Eng Chem Res 47(23):9288–9296
131. Ware RA, Wei J (1985) Catalytic hydrodemetallation of nickel porphyrins: I. Porphyrin structure and reactivity. J Catal 93(1):100–121
132. Ware RA, Wei J (1985) Catalytic hydrodemetallation of nickel porphyrins: III. Acid-base modification of selectivity. J Catal 93(1):135–151
133. Ware RA, Wei J (1985) Catalytic hydrodemetallation of nickel porphyrins: II. Effects of pyridine and of sulfiding. J Catal 93(1):122–134
134. Furimsky E, Massoth FE (1999) Deactivation of hydroprocessing catalysts. Catal Today 52 (4):381–495
135. Scheuerman GL, Johnson DR, Reynolds BE, Bachtel RW, Threlkel RS (1993) Advances in Chevron RDS technology for heavy oil upgrading flexibility. Fuel Process Technol 35 (1):39–54

136. Hung C-W, Wei J (1980) The kinetics of porphyrin hydrodemetallation. 1. Nickel compounds. Ind Eng Chem Process Design Dev 19(2):250–257
137. Gray RM (1994) Upgrading petroleum residues and heavy oils. Marcel Dekker, Inc.:New York
138. Dechaine GP, Gray MR (2010) Chemistry and association of vanadium compounds in heavy oil and bitumen, and implications for their selective removal. Energy Fuel 24(5):2795–2808
139. Gould KA (1980) Oxidative demetallization of petroleum asphaltenes and residua. Fuel 59(10):733–736
140. Dedeles GR, Abe A, Saito K, Asano K, Saito K, Yokota A, Tomita F (2000) Microbial demetallization of crude oil: nickel protoporphyrin disodium as a model organo-metallic substrate. J Biosci Bioeng 90(5):515–521
141. Ovalles C, Rojas I, Acevedo S, Escobar G, Jorge G, Gutierrez LB, Rincon A, Scharifker B (1996) Upgrading of orinoco belt crude oil and its fractions by an electrochemical system in the presence of protonating agents. Fuel Process Technol 48(2):159–172
142. Sakanishi K, Yamashita N, Whitehurst DD, Mochida I (1998) Depolymerization and demetallation treatments of asphaltene in vacuum residue. Catal Today 43(3):241–247
143. Shiraishi Y, Hirai T, Komasawa I (2000) A novel demetalation process for vanadyl-and nickelporphyrins from petroleum residue by photochemical reaction and liquid–liquid extraction. Ind Eng Chem Res 39(5):1345–1355
144. Sakanishi K, Saito I, Watanabe I, Mochida I (2004) Dissolution and demetallation treatment of asphaltene in resid using adsorbent and oil-soluble MO complex. Fuel 83(14):1889–1893

Ruthenium Ion-Catalyzed Oxidation for Petroleum Molecule Structural Features: A Review

Quan Shi, Jiawei Wang, Xibin Zhou, Chunming Xu, Suoqi Zhao, and Keng H. Chung

Abstract Ruthenium ion-catalyzed oxidation (RICO) is an oxidative degradation approach for the structural investigation of petroleum fractions. It is based on the selective oxidation and near quantitative removal of aromatic carbon from aromatic petroleum fractions, while leaving the structural integrity of aliphatic units intact. RICO method has played a highly useful role in the investigation of the molecular structures of heavy petroleum. It distinguishes alkyl groups attached to aromatic rings, alkyl bridges between aromatic rings, the nature of aromatic condensation, etc. The application of RICO to petroleum chemistry was promoted by Strausz and coworkers for the study of asphaltene and other high molecular weight petroleum fractions. Structural details on asphaltenes and their ramifications were revealed by the RICO-based studies, and a hypothetical molecular model was proposed for the petroleum asphaltenes. The structural information obtained from the model is valuable for understanding such complex molecular systems. Another application of the RICO technique in the petroleum industry is the characterization of biomarkers in heavy petroleum fractions and kerogen, which were connected to the condensed core structures by chemical bonds. Generally, only the oxidation of aromatic carbon to carbon dioxide and carbonyl functionalities in RICO is considered; however, other reactions may also take place. Since they occur parallel to the oxidation of aromatic carbon, misinterpretation of the relevant experimental results may result. Recent research based on ultrahigh-resolution mass spectrometry has provided new evidence for the side reactions, which leads to a more informative

---

Q. Shi (✉), J. Wang, C. Xu, and S. Zhao
State Key Laboratory of Heavy Oil Processing, China University of Petroleum, Beijing 102249, China
e-mail: sq@cup.edu.cn

X. Zhou
College of Basic Science, Liaoning Medical University, Jinzhou, Liaoning 121001, China

K.H. Chung
Liaoning Huajin Tongda Chemicals Co. Ltd., Panjin, Liaoning 124000, China

interpretation of the RICO results. This paper reviews the RICO-related studies on petroleum fractions. The interpretation of RICO experimental results is also discussed.

Keywords Asphaltene · Heavy oil · RICO

Contents

1 Introduction 72
2 Principles and Methodology 74
    2.1 Mechanism of RICO Catalytic Reaction 74
    2.2 Basic Reactions of RICO 74
    2.3 Other Side Reactions Occurring in RICO System 77
    2.4 Quantitative Analysis of RICO Products 78
3 RICO Revealed Molecular Structure of Heavy Petroleum Fractions 80
    3.1 Asphaltenes 80
    3.2 Other Heavy Petroleum Fractions 83
    3.3 Carbon Residue on Catalyst 84
    3.4 Kerogen 85
4 Summary and Future Prospects 86
References 87

1 Introduction

Petroleum is an extremely complex mixture of hydrocarbon with heteroatoms. The research on molecular composition and structure of petroleum systems is challenging. Many questions and disputes have arisen in the petroleum chemistry community regarding the molecular structures of heavy petroleum fractions, especially the asphaltenes. The two key barriers in characterization of heavy petroleum fractions are the difficulty of separating the complex mixture into individual compounds and the complexity and dispersity of heavy petroleum systems, which are beyond the detection capability of advanced analytical techniques [1, 2].

Ruthenium ion-catalyzed oxidation (RICO) has been a useful sample preparation technique for investigating the molecular structures of aromatic units in heavy petroleum fractions [3]. RICO converts un-substituted aromatic carbons to carbon dioxide without altering the structural integrity of aliphatic and naphthenic units. This allows the estimation of the side chain distribution of alkyl groups which are attached to the aromatic carbons and the methylene bridges connecting two aromatic carbons in macromolecules of heavy petroleum fractions.

RICO removes almost all aromatic carbons in the petroleum system and keeps the aliphatic and naphthenic structures unchanged. Attempts have been made by using the RICO technique to quantify the various molecular structures in heavy petroleum fractions [4]: (1) the distribution of n-alkyl chains attached to aromatic carbons, i.e., chain length, total number of n-alkyl chains, and the number of carbon atoms; (2) the distribution of polymethylene bridges connecting two aromatic units, i.e., bridge length, total number of bridges, and the number of carbon atoms; and (3) the amount of saturated sulfides. RICO can also be used to estimate semiquantitatively the carbon atoms in bridges connecting an aromatic attached to a naphthenic ring and that of the aliphatic carbon attached to naphthenic rings. Various isomeric carboxylic acids of benzenemono- to hexa-types, which are produced from the condensed aromatic nuclei during the RICO reactions, provide some insights into the degree of aromatic condensation of the asphaltene molecules [5–11]. RICO can also be used to determine the number of *n*-alkanoic acid esters anchored to the aliphatic/naphthenic cores of the asphaltene molecules [4].

RICO was first introduced by Djerassi and Engle [12] in 1953, but was not widely accepted by organic chemists, due to the high variability and inconsistency of the experimental results. In 1981, Carlsen et al. [13] used acetonitrile as a co-solvent for RICO reactions. Acetonitrile prevents the precipitation of lower-valent ruthenium carboxylate complexes; hence it improved the reproducibility, selectivity, and efficiency of RICO oxidation reactions [4].

RICO was first applied to the fossil fuel system by Stock and coworkers [14] in 1983, followed by Mallya and Zingaro [15] in 1984 for the study of coal molecular structures. Ruthenium tetraoxide oxidation reactions were used to convert coal macromolecules into a wide range of aliphatic carboxylic and benzenecarboxylic acids. This provided qualitative and quantitative information on aliphatic and aromatic structural elements in coal [14–22] and allowed the researchers to postulate a block structural diagram of coal using the principal concepts [21]. The results showed that coal has a wide structural diversity, which cannot be adequately elucidated by any conventional molecular structure representation that meets the requirement for structural completeness in the traditional sense of the natural product chemistry [21]. Nomura et al. improved the validity of the RICO technique [23] and applied it to determine the aliphatic bridges and functional groups in various coal samples [24–28] and petroleum fractions [29–31]. RICO has been one of the most effective approaches for the coal structure investigation [32–45].

In 1985, Strausz et al. applied RICO to study of asphaltenes and heavy petroleum fractions [4, 6–8, 46–48]. Structural details of asphaltenes and their ramifications were revealed by the RICO-based studies [7, 8]. Strausz et al. conducted an extensive study on Alberta oil sand bitumen and heavy oils and proposed a hypothetical petroleum-derived asphaltene molecular model [47]. Although the proposed asphaltene model has been controversial [49–57], some aspects of the model structure information are valuable for the understanding of a highly complex molecular system. Another application of RICO technique in petroleum system is the characterization of biomarkers in heavy petroleum fractions and kerogen, which connect the condensed core structures by chemical bonds [45, 48, 58–65].

2 Principles and Methodology

2.1 Mechanism of RICO Catalytic Reaction

Early literature referred RICO as "ruthenium tetroxide catalyzed" or "Ru(VIII) catalyzed" reactions which are not only ambiguous designations, but are technically incorrect [4]. Ruthenium tetroxide or Ru(VIII) is only one of the oxidative states in the catalytic cycle:

Ru(III)-NaIO$_4 \rightarrow$Ru(VIII)-organic reducing agent $\rightarrow$ Ru(III)/Ru(II)-NaIO$_4 \rightarrow$Ru (VIII)

Ru(VIII) (RuO4) is the actual oxidant for the reactions and is a catalyst in the reaction cycle. The RuO$_2$ of RuCl$_3$ is the oxidation agent for RICO, while periodate or hypochlorite is the stoichiometric oxidant. In early studies, the RICO reactions were slow and/or incomplete. Hence, the early RICO reaction results were time dependent and not repeatable and not widely accepted until 1981 when Sharpless suggested the use of acetonitrile as a co-solvent for RICO reactions [13]. Acetonitrile prevents the precipitation of lower-valent ruthenium carboxylate complexes, which improved the reproducibility, selectivity, and efficiency of RICO oxidation reactions [4]. Solvent composition is a critical parameter for RICO reactions; two suggested ratios of CCl4:CH3CN:H2O were popular: 2:2:3 and 1:1:2 (in volume); the former was suggested as the most optimized solvent ratio by Carlsen et al. [13].

2.2 Basic Reactions of RICO

An extensive series of model compounds have been studied for the RICO reaction over the past three decades, and the results are summarized in Table 1. From these studies, the RICO reaction mechanisms can be depicted as shown in Fig. 1 [3, 4].

In general, the RICO reaction mechanisms are based on the hypothesis that sp^2 aromatic carbons are selectively oxidized by RICO agents. Alkyl-substituted aromatics are oxidized to produce CO_2, except at the alkyl attachment site which is converted to a carboxylic group attached to the alkyl chain, resulting to the formation of an alkanoic acid. For an aliphatic chain located between two aromatic carbons (in one or two aromatic rings), it produces a dicarboxylic acid and CO_2. For fused ring aromatic molecules, some aromatic carbons are oxidized to form carboxylic groups which attached to a benzene ring. The carboxylic groups deter further oxidation of the benzene ring, leading to the formation of a third group of acids: a suite of benzenedi-, tri-, tetra-, penta-, and hexacarboxylic acids [29].

The pattern of reactivity is a relatively simple one from which rather secure structural information can be obtained [14]. Alkanoic acids are derived from alkyl substituted groups attached to the aromatic cores. Dicarboxylic acid can be derived from polyalkylene bridge or cycloparaffinic ring combined with aromatic ring. Benzenepolycarboxylic acids are derived from condensed aromatic structures in

Table 1 The results of model compounds subjected to RICO

Model compound	Product distribution/%	Conversion/%	Ref
Benzene	Vigorous explosion		[12]
Phenanthrene	Phenanthraquinone	28	[12]
Diphenyl sulfide	Diphenyl sulfone	31	[12]
Methyl benzyl sulfide	Methyl benzyl sulfone	58	[12]
Pyridylphenylmethane	4'-Pyridylphenone	67	[66]
Cyclohexylbenzyl ether	Cyclohexyl benzoate	90	[66]
Propylbenzene	Butyric acid (95); propiophenone (5)	76	[14]
Tridecylbenzene	Tetradecanoic acid (91); tridecanophenone (9)	70	[14]
Tetralin	Adipic acid (75); 1-tetralone (8); glutaric acid (17)	100	[14]
Bibenzyl	Succinic acid (35); hydrocinnamic acid (63)		[14]
4-Pentylbiphenyl	Hexanoic acid (51); benzoic acid (54); 4-pentylbenzoic acid (38)	74	[14]
Phenanthrene	Phthalic acid (5); diphenic acid (91); phenanthrenequinone (4)	100	[14]
Triphenylene	Phthalic acid (75); benzenehexacarboxylic acid (25)	55	[4]
9,10-Dihydrophenanthrene	Succinic acid (2); phthalic acid (7); diphenic acid (55); 3-(2-carboxyphenyl)propionic acid (32)	81	[14]
1-Decene	Nonanoic acid		[13]
Phenylcyclohexane	Cyclohexanecarboxylic acid	89	[13]
2,2-Diphenylpropane	Dimethylmalonic acid	–	[22]
6-Hydroxytetralin	Glutaric acid; succinic acid	–	[22]
6-Methoxytetralin	Glutaric acid; succinic acid	–	[22]
2-Tetralone	Succinic acid	–	[22]
Lignite coals	Phthalic and other polybenzenecarboxylic acids	–	[67]
Diheptyl sulfide	Diheptyl sulfone	99	[4]
Dibenzyl sulfide	Dibenzyl sulfone	88	[4]
Diphenyl sulfide	Diphenyl sulfone	85	[4]
Dibenzothiophene	Dibenzothiophene sulfone	~100	[4]
n-Alkyl thiolanes and thianes	Thiolane and thiane sulfones	~100	[4]
Benzothiophene	Completely oxidized	–	[4]
1-Hexadecene	Pentadecanoic acid	93	[4]
1,12-Diphenyl dodecane	1,12-tetradecanecarboxylic acid	100	[4]
Hexadecanoic acid	Hexadecanoic acid	–	[9]
Butylbenzene	Valeric acid (79.5); butyric acid (4.1); propionic acid (1.0); acetic acid (0.5)	–	[9]

(continued)

Table 1 (continued)

Model compound	Product distribution/%	Conversion/%	Ref
Dodecylbenzene	Tridecanoic acid (72.0); dodecanoic acid (3.5); undecanoic acid ~ nonanoic acid (0.5); dodecanophenone (8.9); benzoic acid (0.3)	85.2	[9]
1,6-Diphenylhexane	Suberic acid (56.1); pimelic acid (4.6); adipic acid (1.2); glutaric acid (0.6); benzoic acid (7.3); 6-benzoyl hexanoic acid (15.5); other products (14.7)	100	[9]
Naphthalene	Phthalic acid (100)	100	[9]
Tetralin	Adipic acid (62.6); glutaric acid (15.7); 1-tetralone (21.7)	100	[9]
Anthracene	Phthalic acid (56.4); anthraquinone (24.0); pyromellitic acid (19.6)	100	[9]
Phenanthrene	Phthalic acid (59.1); 1,1'-diphenic acid (40.9)	100	[9]
Biphenyl	Benzoic acid (91.7)	98.6	[9]
Diphenyl methane	Benzoic acid (12.3); phenylacetic acid (48.3); diphenyl ketone (26.3)	98.4	[9]
n-Octane	Octanone (99); octanol (<1)	5	[68]
n-Hexacosane	Hexacosanones (99); hexacosanols (<1)	24	[68]
n-Triacontane	Triacontanones (99); triacontanols (<1)	36	[68]
n-Dotriacontane	Dotriacontanones (99); dotriacontanols (<1)	36	[68]
Cycloheptane	Cycloheptanone (99); cycloheptanol (<1)	40	[68]
Methylcyclohexane	Methylcyclohexanones (27); methylcyclohexanol (72)	13	[68]
n-Dodecylcyclohexane	Dodecylcyclohexanone (34); dodecylcyclohexanols (61)	27	[68]
n-Nonadecylcyclohexane	Nonadecylcyclohexanone (38); nonadecylcyclohexanols(-)	38	[68]
Bicyclohexyl	Bicyclohexanones (35); bicyclohexanol (62)	30	[68]
Bicyclo(4,4,0)decane	Bicyclo(4,4,0)decanone (17); bicyclo(4,4,0)decanol (78)	51	[68]
Adamantane	Adamantanones (4); adamantanols (93)	91	[68]
Squalane	Squalanones (20); squalanols (76)	53	[68]
Cholestane	Cholestanones (58); cholestanols (24)	39	[68]

the asphaltenes [14, 47]. The formation of benzenepolycarboxylic acids such as 1,2,4-benzenetricarboxylic acid and 1,2,3,5-benzenetetracarboxylic acid indicates the presence of biphenyl-type aromatic structure in the molecule. The formation of phthalic acid and 1,2,3,4-benzenetetracarboxylic acid indicates the presence of aromatic structure with combined naphthalene and combined anthracene types in the molecule. The formation of 1,2,3-benzenetricarboxylic acid, 1,2,3,4-benzenetetracarboxylic acid, benzenepentacarboxylic acid, and benzenehexacarboxylic acid indicates the presence of highly condensed aromatic nuclei in the molecule [5, 47], an indication of a large aromatic molecular system.

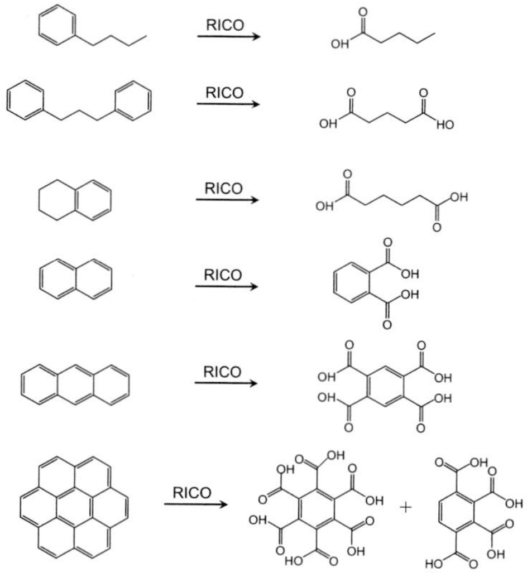

Fig. 1 Typical RICO reactions for aromatic hydrocarbons [3, 4]

2.3 Other Side Reactions Occurring in RICO System

The basic assumption used to describe the formation of aromatic RICO reaction products is that the aromatic carbons are oxidized to form carbon dioxide and carbonyl functionalities. In reality, there are also side reactions taking place at slower rates than the main RICO reactions [7, 69, 70]. RuO_4 is known to oxidize alcohols, aldehydes, alkenes, alkynes, and some aromatic compounds to form carboxylic acids [71], secondary alcohols to ketones, ethers to esters, and sulfides to sulfones [7]. In addition, aliphatic C–H bonds may also be oxidized to form alcohols or ketones [68, 72, 73]. These side reactions occurring in the RICO system have not been fully discussed or overlooked in previous RICO studies [74].

Acetonitrile, the co-solvent used for RICO reactions, improves the oxidation reaction efficiency. However, acetonitrile itself can be hydrolyzed and/or oxidized during the RICO reactions, leading to the formation of acetic acid which is simultaneously produced in the oxidation of methyl aromatic compounds [7]. The reaction activity of acetonitrile increases at low pH in the presence of carboxylic acids which are produced in the RICO reactions. Hence, the acetonitrile reaction products could result in overestimation of methyl groups attached on the aromatic units [7].

Tertiary C–H bonds at the bridgehead positions can be oxidized by RICO to form alcohols and the secondary C–H bonds to ketones [72]. Based on this side reaction, Zhou et al. [68] developed a novel technique to characterize the saturated fraction of heavy petroleum fractions. The oxidized product alcohols were analyzed by negative ion electrospray Fourier transform ion cyclotron resonance mass

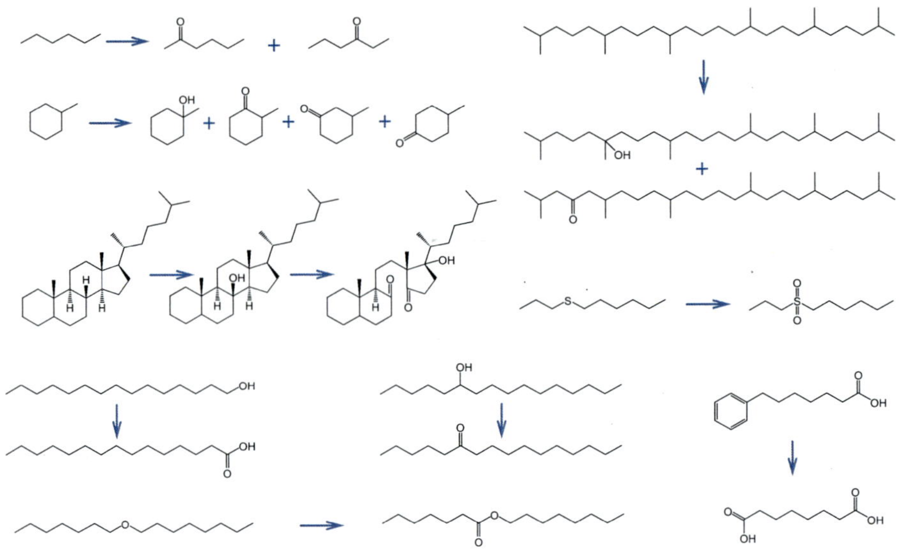

Fig. 2 Some reactions can occur in the RICO process [7, 73, 75]

spectrometry (FT-ICR MS). Subsequently, Zhou et al. [74] used this technique to characterize the molecular composition of saturate fractions of six heavy petroleum fractions.

Primary alcohols attached to aromatic carbons can be oxidized to form α,-ω-di-alkanoic acids which could also be produced from aromatic-attached fatty acids. In addition, ethers can be oxidized to form esters; hence the presence of methyl ethers in RICO reaction products could be misinterpreted as carboxylic groups in the molecule. These side reactions could result in formation of dicarboxylic acids, increasing the yield of this series arising from the oxidation of alkyl bridges connecting two aromatic units.

Other side reactions include transformation of sulfides to sulfones and perhaps sulfinic and sulfonic acid esters. Fluorenes or other hydroaromatics can be oxidized to keto acid [7]. Figure 2 presents some possible reaction pathway of the RICO for petroleum.

2.4 Quantitative Analysis of RICO Products

Ruthenium (VIII) catalyzes oxidation reactions of aromatic components in fossil fuels to produce a mixture of aliphatic and aromatic carboxylic acids. The acids are further converted into methyl esters and subjected to gas chromatogram–mass spectrometry (GC-MS) analysis or other gas chromatographic techniques, such as GC Fourier transform infrared spectroscopy (GC-FTIR) [18]. However, the deficiency of RICO reaction system is the lack of quantitative analysis of carboxylic acids which are the main reaction product. Other reaction products include CO_2, homologous series of normal alkanoic acids (C_2–C_{35}), α,ω-dicarboxylic acids, and a suite of benzenedi-, tri-, tetra-, penta-, and hexacarboxylic acids. Quantitative analysis of CO_2 produced in RICO system can be carried out by one of the two techniques: CO_2 absorption by Ascarite and determining the weight change of Ascarite before and after the experiment [76] and CO_2 absorption by $Ba(OH)_2$ followed by titration with HCl solution [8].

Carboxylic acids are required to be esterified for GC analysis. The difficulty associated with quantifying RICO reaction system is the loss of low molecular weight, volatile reaction products [77]. Attempts have been made to capture and analyze the volatile RICO reaction products. Stock and Wang [16] used the isotope dilution technique to determine the amount of the volatile monocarboxylic acids with two to six carbon atoms. Peng et al. [6] use a three-step esterification of RICO reaction products by reacting *n*-alcohols of various chain lengths—methanol, CH_2N_2 to form methyl, *n*-octyl, and *n*-octadecyl alcohols followed by octyl and octadecyl esters, respectively. The product alkanoic acids up to C_{12} acids were octadecyl esters; diacids up to C_{12} were dioctyl esters; all other acid products were methyl esters. Esterification was done on the acid-derived sodium salts in the presence of boron tribromide. The optimum yield of the esters (of the C_2, C_4, C_8, and C_{11} *n*-acids and succinic acid (C4)) was 90%. Murata et al. [24] used ion chromatography to quantify low molecular weight carboxylic acids in RICO reaction products by minimizing the loss of highly volatile matters.

The analysis of acids in the aqueous phase is also challenging for the RICO system. Most of the benzenepolycarboxylic acids and C_4–C_8 dicarboxylic acids of RICO reaction products are in the aqueous phase. Mojelsky et al. [4] performed direct treatment of the aqueous phase of RICO reaction products with an excess amount of diazomethane. However, esterification efficiency of the RICO products was low, due to rapid decomposition of diazomethane when it is in contact with acidic water solution. This could explain the low concentration of benzenepentacarboxylic and benzenehexacarboxylic acids found in RICO reaction products, once leading to the conclusion that "pericondensed aromatic units play a very minor role in the molecular structure of petroleum asphaltene" [7].

Wang et al. [5, 9, 10, 77] performed dewatering of RICO reaction product prior to methylation. The aqueous phase was vacuum distilled at 90°C, and the solid residue was treated with an excess amount of ethereal diazomethane. After methylation, the solid residue was extracted with dichloromethane to obtain the esters. Model compound study showed that benzenepolycarboxylic acids dissolved in water can be quantitatively recovered; the conversion efficiency of benzenehexacarboxylic was improved from 17% to 98% [77].

Based on the success of quantitative analysis for benzenepolycarboxylic acids, Wang et al. [5] proposed the PCI to define the degree of condensed aromatic units in heavy petroleum fractions, as shown in Eq. (1):

$$PCI = \frac{PERI}{CATA}, \qquad (1)$$

where PERI is the amount of 1,2,3-tri- + 1,2,3,4-tetra- + 1,2,3,5-tetra- + penta- + hexa-benzenecarboxylic acids and CATA is the amount of di- + 1,2,4-tri- + 1,2,4,5-tetra-benzenecarboxylic acids. The PCI values of aromatics, resins, and asphaltenes of Shengli vacuum residue were 0.64, 1.4, and 2.3, respectively [5].

BIPH is a parameter to define the aromatic units of biphenyl structure:

$$BIPH = 1,2,4\text{-tri-} + 1,2,3,5\text{-tetra-benzenecarboxylicacid}. \qquad (2)$$

3 RICO Revealed Molecular Structure of Heavy Petroleum Fractions

3.1 *Asphaltenes*

For petroleum chemists, "asphaltenes" are generally applied in the narrow sense of petroleum asphaltene [7]. Asphaltenes are the "heaviest" fraction of petroleum, which has the highest molecular weight (MW), N, S, O, and metal content and the lowest H/C ratio among the petroleum fractions. Asphaltenes are known to exhibit an intractable and undecipherable molecular structure.

Tremendous progress on the chemistry of heavy petroleum fractions was made by Strausz and his coworkers, who used RICO as a primary technique to investigate the molecular structure of asphaltenes [4, 6–8, 46, 47]. In 1984, Strausz et al. [76] applied the RICO reaction system to study the chemistry of heavy petroleum fractions. They developed experimental protocols to delineate the molecular transformation of heavy petroleum fractions during RICO reactions. Various sampling techniques to recover the RICO reaction products were reported in 1996, which allowed more quantitative and accurate interpretation of experimental data to elucidate the molecular structure transformation of heavy petroleum fractions [6, 8].

Mojelsky et al. [4] proposed the molecular structures of asphaltenes from Alberta oil sand bitumen and heavy oil. Figure 3a, b show the distribution of n-alkyl groups and methylene chains, respectively, attached to aromatic moieties of asphaltenes.

Based on the RICO results of Athabasca asphaltene, the estimated aliphatic and aromatic carbon contents were 26.5 and 34.6% of total carbon, respectively [4]. The

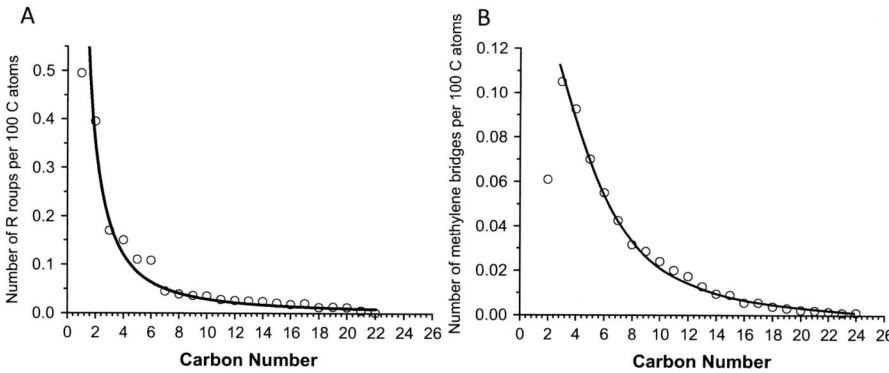

Fig. 3 Distribution of *n*-alkyl and the methylene chains attached to aromatic moieties in Athabasca asphaltenes (modified from reference [4]). (**a**): *n*-Alkyl groups; (**b**): methylene chains, which correspond to the alkyl bridges between two aromatic units

aliphatic carbon content determined by RICO was in agreement with that by NMR analysis. The aromatic carbon content determined by RICO was 80% of total aromatic carbon originally present in asphaltene. The missing 20% attributed to losses due to volatility, occurrence of side reactions, appearance of unidentified products, inherent errors, and decarboxylation of the carboxylic acids [4]. The authors stated that all the yield values were lower limits owing to losses due to volatility; however, a more plausible explanation is the lower recovery of esterification for polybenzenecarboxylic acids. Their GC analysis did not show the presence of benzenepenta- and benzenehexacarboxylic acids in the oxidized products. In their subsequent study, Strausz et al. [47] indicated the presence of small quantities of pentacarboxylic and hexacarboxylic acids which were 4.3 and 1.5 mol % of the total benzenepolycarboxylic acid yield. Wang et al. [76] performed the model compound study of RICO and indicated a low yield (17%) of benzenehexacarboxylic acid from esterification reactions; however, the yield was increased to 98% with a modified process of methylation.

Mojelsky et al. [4] performed a comparative study of asphaltenes from various crude oils obtained from the Western Canadian sedimentary basin: Peace River bitumen, Peace River steam-produced bitumen, Carbonate Triangle bitumen, and Lloydminster heavy oil. The results showed that all five asphaltene samples had a similar molecular structure. Small deviations among the asphaltene samples were attributed to slight variations in reservoir conditions, i.e., temperature, chemically active minerals, water washing, activity of sulfate-reducing bacteria, and other microbiological processes. These asphaltenes are likely originally derived from *n*-alkane-derived kerogens.

Strausz et al. [8] reviewed the RICO experimental procedure and results and proposed a hypothetical molecular model of asphaltenes, comprising an elemental formula of $C_{420}H_{496}N_6S_{14}O_4V$, an H/C atomic ratio of 1.18, and a molecular weight of 6,191 Da. The C, H, S, N, O, and V contents are 81, 8, 7.3, 1.4, 1.0, and 0.8, respectively.

The proposed asphaltene molecule exhibited a loose, flexible architecture in which aromatic and naphthenic ring structural units were attached to the asphaltene core by $-(CH2)n-$, $-S-$ and possibly alkyl ester $-C(O)-O-$ and $-O-C(O)-$ and alkyl ether $-O-$ bridges. Along with bridges, these structural elements were also present as side chains.

By incorporating pyrolysis and nickel boride reduction studies, some structural features of asphaltene molecule were described and are summarized below [74]:

- n-Alkyl side chains from RICO reactions were methyl esters up to C_{30}.
- α-C_1–C_4 branched n-alkyl side chains from RuVIII oxidation were methyl esters up to C_{31}.
- α,ω-di-n-alkyl bridges from RICO reactions were dimethyl esters.
- n-Alkyl- and monomethyl-n-alkylbenzene were up to C_{28}.
- Aromatic condensed naphthenic rings were usually with alkyl side chains.
- n-Alkylthiophenes were up to C_{29}.
- n-Alkylthiolanes and thianes were up to C_{29}.
- n-Alkylbenzothiophenes were up to C_{24}.
- n-Alkyldibenzothiophenes were up to C_{25}.
- Some other condensed n-alkylthiophenes were up to C_{26}.
- 9-n-Alkylfluorenes and various biological markers were attached directly to aromatic carbon in the asphaltene core or via one or two sulfur atoms.

Compared to the results obtained from mass spectrometry analysis and other techniques, the proposed asphaltene model vastly overestimated its molecular weight which was determined by vapor pressure osmosis (VPO). It has been shown that the asphaltene molecules form aggregates in the VPO solvent system [77–82]. Hence, the macrostructures of the proposed asphaltene model were erroneous; nevertheless, most of the microstructures were still relevant and useful for understanding the asphaltene chemistry.

Wang et al. [77] modified the esterification of RICO aqueous phase products, leading to increased recovery of benzenepolycarboxylic acids. The oxidized products were dominated by a homologous series of straight-chain monocarboxylic acids (C_2–C_{35}), indicating that normal alkyl chains are important constituents of the asphaltenes. A series of α,ω-dicarboxylic acids (C_4–C_{26}) were also detected and determined, validating the presence of polymethylene moieties linking two aromatic units. Relative high yields of benzenepenta- and hexacarboxylic acids formed in the RICO reactions of Gudao asphaltenes, indicating that substantial aromatic structures of asphaltene are peri-condensed [77]. The conclusion was "corrected" by Strausz: a peri-condensed aromatic structure in asphaltene chemistry is meant to denote a condensed aromatic sheet in which at least one of the rings is completely surrounded by aromatic rings, while Wang et al. use an organic chemical definition in which three rings share one common aromatic carbon [7]. Regardless, Gudao asphaltenes did exhibit a more condensed aromatic structure than Athabasca asphaltenes reported by Strausz et al. [8, 77]. The finding of asphaltenes containing highly condensed aromatic units was further confirmed by many studies [3, 9–11, 78–83] published in Chinese literature and on asphaltenes derived from various Chinese crudes: Shengli, Liaohe, Gudao, Daqing, Weizhou, Lunnan, and Tahe [9–11, 59, 60, 62, 77, 81–87]. However, Zhu et al. [11] found benzoic acid and *p*-benzenedicarboxylic acid in the oxidized products, suggesting that biphenyl-like aromatic units are present in asphaltenes, which has been overlooked. Zhu et al. [11] calculated the structural parameters of aromatic rings in four asphaltene samples and concluded that the structural parameters were dependent on the oil maturity. The PCI for Kd45, P14, T401, and Zh7 asphaltenes were 0.526, 0.423, 0.644, and 0.199, respectively, which were much lower than those for Shengli asphaltenes of 2.3 reported by Wang et al. [5].

3.2 Other Heavy Petroleum Fractions

RICO was also used to study the molecular structures of vacuum residues. Wang et al. [9] applied RICO reactions to aromatic and resin fractions of Daqing residue. The C_2–C_{31} *n*-alkyl acids accounted for more than 50 wt% of total acids in the oxidized products. The range of carbon number distribution for the aromatics was similar to that for the resins, except that the aromatics had more alkyl chains. Very low quantities of C_4–C_{13} diacids were detected in the product (an order of magnitude lower than *n*-alkyl acids), indicating the presence of alkyl bridges in the maltenes.

Wang et al. [10] performed a comparative molecular structure study on five residues. The resins were separated by varying polarity into three subfractions: light, middle, and heavy resins. Each resin subfraction and asphaltenes were subjected to RICO reactions. The distribution range of *n*-alkyl groups attached to the aromatic moieties was similar to that of methylene bridge groups connected to two aromatic moieties. The relative abundances of carbon number and the content of total alkyl chains varied among resin subfractions. In general, the results indicated that there were no structural gaps among various resin subfractions and asphaltenes of various residua.

Zhang et al. [79, 83] investigated the structural changes of an Omani vacuum residue and its subfractions during hydroprocessing in a five-stacked catalyst bed reactor system. The reaction product from each stack bed was obtained: hydrodemetallization product 1 (HDM1), hydrodemetallization product 2 (HDM2), hydrodesulfurization product 1 (HDS1), hydrodenitrogenation product 1 (HDN1), and hydrodenitrogenation product 2 (HDN2). The residue and its five hydroprocessed products and their subfractions (aromatics, resins, and asphaltenes) were subjected to RICO reactions. All the residue subfractions had abundant C_1–C_{27} alkyl side chains, mostly *n*-alkyl chains. The aromatic structure of aromatic fraction was mostly cata-condensed, and that of resin fraction had comparable amounts of cata-condensed and peri-condensed aromatics, whereas that of asphaltene fraction was mostly peri-condensed. The change in chemical structure of resid subfractions was dependent on the degree of hydrotreatment. Asphaltene fraction had more and highly condensed aromatic structures due to its enriched large aromatic units.

3.3 Carbon Residue on Catalyst

Hydrotreating is an inevitable and effective refinery process for removing contaminants (such as sulfur, nitrogen, and metals) in heavy petroleum fractions. A challenging operating issue related to catalysis of heavy petroleum fractions is rapid catalyst deactivation due to carbon deposition on catalyst. It is believed that a complex mixture of condensed polyaromatic compound deposits on catalyst (catalyst surface, pores, and channels) as a reaction by-product which further evolves to form coke, a bulky condensed insoluble organic matter under severe process conditions. From an analytical perspective, the structure and composition characterization of soluble coke precursors can be performed using various analytical techniques, but that of coke (insoluble matter) is still lacking.

Zhang et al. [79, 83] investigated the structure of coke on catalyst for five spent residue hydroprocessing catalysts using the RICO reaction system. The distribution of n-alkyl chains and bridges attached to aromatic carbon of coke on catalysts was investigated. The substitute group connected with aromatic core structures was mainly methyl and no $> C_5$ alkyl chains in coke. Various isomeric benzenepolycarboxylic acids in RICO oxidized products were identified. The condensation mode of some aromatic structure was discussed. The structure of coke formed was dependent on the coke precursor and its reaction pathways. Catalysts promoted coke dehydrogenation and aromatic condensation reaction, leading to formation of graphitic coke (60% of total coke yield).

An intriguing aspect of Zhang et al.'s work was the finding of α,ω-dicarboxylic acids present in coke RICO reaction products, which were postulated to derive mainly from $-CH_2CH_2-$ in coke and some from $-CH_2CH_2CH_2-$ and $-CH_2CH_2CH_2CH_2-$, which were alkyl bridges attached to two aromatic units or cycloparaffinic rings combined with aromatic rings. However, by combining the compositions of all other products, the most likely logical explanation for the formation of dicarboxylic acids is from oxidized naphthenic ring in the partial saturated aromatics, based on the presence of (1) abundant 1,4-butanedioic acid, 1,5-pentanedioic acid, and 1,6-hexanedioic acid, which were the only dicarboxylic acids detected in the RICO products; (2) increased abundance of these compounds in deep hydroprocessed products; and (3) the longest alkyl chain of fatty acids which was C_5 [88].

3.4 Kerogen

Kerogen is considered as the petroleum precursor; hence, the molecular structure of kerogen is likely to resemble that of heavy petroleum. Due to the heterogeneity and complexity of kerogen, selective chemical degradations with various reagents were employed to fragment a large kerogen molecule into various smaller units in order to gain insights into the types of moieties and the nature of the bonds linking them to form the macromolecular matrix [64].

RICO reaction system has been used for structural investigation of kerogen molecules [40, 63, 64, 71, 89–97]. The degradation products from kerogen were similar to those from petroleum asphaltenes [62], comprising normal and branched α,ω-dicarboxylic acids, normal and branched monocarboxylic acids, tricyclic terpenoid acids, hopanoic acids, gammacerane carboxylic acids, pregnanoic acids, sterane carboxylic acids, and benzenemono- to hexa-carboxylic acids. However, the amounts of these compounds varied significantly from sample to sample, indicating that kerogen has a wide range of molecule compositions. Eglinton and coworkers [95, 96] subjected a variety of type I, II, and III kerogens to RICO and found that, in all cases, straight-chain acids were the predominant products. However, other studies reported a different composition of the kerogen RICO product: α,ω-dicarboxylic acids were more abundant than monocarboxylic acids [63, 67, 97].

Boucher et al. investigated a type II clay kerogen by a variety of chemical techniques. The dichloromethane-soluble extract of the kerogen RICO product contained a series of straight-chain monocarboxylic acids (14%) in the C_9–C_{27} range, α,ω-dicarboxylic acids (51%) in the C_4–C_{27} range, several branched chain mono- and diacids (9%), and relatively large amounts of C_{14}–C_{17} and C_{19}–C_{21} acyclic isoprenoid acids (16%). The RICO and ^{13}C-NMR spectroscopy provided complementary compositional information on aliphatic and aromatic components, respectively. The finding was that the carbon skeletons of the kerogen were mainly linear aliphatic chains with relatively small amounts of cyclic structures.

Messel shale kerogen was readily used in many RICO studies. Standen et al. [98], Reiss et al. [94], Blokker et al. [99], and Dragojlović et al. [71] reported that the sources of α,ω-dicarboxylic acids (C_{10}–C_{13}, with a maximum at C_{10}) in the kerogen RICO product could have been α,ω-disubstituted or α,ω-dienic aliphatic chains in the kerogen matrix and/or originated from oxidation of Cn–2 polymethylene bridges linking aromatic structural units. However, Barakat et al. [89] suggested that α,ω-dicarboxylic acids could be derived from diterminal oxidative cleavage of functionalized polymethylene chains within the kerogen (e.g., polymethylene chains linked to the kerogen matrix by ether groups) or from oxidation of functionalized aliphatic moieties (e.g., n-alkan-1-ols, aldehydes, carboxylic acids) linked to the kerogen structure and not removed by $Na_2Cr_2O_7$ oxidation [89].

Yoshioka et al. [91] performed RICO on marine and lacustrine kerogens and indicated that short chain (C_2–C_5) α,ω-dicarboxylic acids were major compounds in the kerogen RICO products. They suggested that the precursor molecules of α,ω-dicarboxylic acids were melanoidins or unsaturated fatty acids. The maximum C_9 α,ω-dicarboxylic acids were identified only in the oxidized product from the lacustrine kerogen. It could be attributed to the difference between marine and lacustrine kerogens with respect to the precursor of unsaturated fatty acids.

Kribii et al. [93] and Dragojlović et al. [71] had a different view in the interpretation of the RICO results. They suggested that alkyl–aryl ether groups were present in the kerogen, probably as a result of O-alkylation between phenols and aliphatic precursors, which was supported by the RICO studies of Barakat et al. [64, 89] which showed a predominant homologous series of 2-methyl n-alkanoic acid methyl esters. Barakat et al. [64, 89] performed two successive degradation experiments, comprising $Na_2Cr_2O_7$/AcOH and RICO on a Type II-S kerogen after pre-saponification of ester bonds by KOH/MeOH treatment. They reported the presence of a homologous series of 2-methyl-n-alkanoic acids as the prominent series of carboxylic acids in the kerogen RICO product. Their results supported the high specificity of RICO reaction system used in revealing the structural information of alkyl chains substituting the aromatic moieties in the kerogen structure and the presence of aryl–alkyl ether groups in the kerogen [64].

4 Summary and Future Prospects

As a widely accepted method for molecular composition investigation, RICO has promoted the understanding of chemical structures of petroleum, coal, and kerogen. It provided significant compositional information that cannot be directly obtained by modern advanced analytical instruments. However, RICO still, to some extent, is an immature method: the experimental process is tedious in step ???, laborious, and time consuming. These features lead to the RICO method having poor repeatability and reproducibility.

Almost all the previous studies have used gas chromatography as the sole approach for reaction product characterization. However, GC is not competent for the complete compositional analysis, since not all components could be eluted through the GC column. Furthermore, unresolved peaks in the GC chromatogram traces usually account for a large portion of the total area, which means that some components in RICO products were not identified and quantified. In this case, extra compositional information should be revealed by in-depth analysis of the reaction products.

Fourier transform ion cyclotron resonance mass spectrometry (FT-ICR MS) has become one powerful tool for the heavy petroleum molecular composition analysis in the past decade [1, 100–102]. Coupled with various ionization sources, ultrahigh-resolution FT-ICR MS could completely resolve the mass peaks of petroleum fractions and unambiguously assign their molecular compositions. Primary studies based on FT-ICR MS have shown that the composition of RICO products was much complicated than that obtained by GC [43, 75]. Future studies should consider combining traditional RICO methods with FT-ICR MS techniques for detailed compositional characterization on large molecules and heteroatom oxidation products to complete the RICO methodology.

References

1. Rodgers RP, Marshall AG (2007) Petroleomics advanced characterization of petroleum-derived materials by Fourier transform ion cyclotron resonance mass spectrometry (Ft-Icr Ms). In: Mullins OC, Sheu EY, Hammani A, Marshall AG (eds) Asphaltenes, heavy oils, and petroleomics. Springer, New York, pp 63–93
2. Hsu CS (2012) Mass resolving power requirement for molecular formula determination of fossil oils. Energy Fuel 26(2):1169–1177
3. Zhang ZG, Guo S, Zhao S, Yan G, Song L, Chen L (2008) Alkyl side chains connected to aromatic units in Dagang vacuum residue and its supercritical fluid extraction and fractions (Sfefs). Energy Fuel 23(1):374–385
4. Mojelsky T, Ignasiak T, Frakman Z, Mcintyre D, Lown E, Montgomery D, Strausz O (1992) Structural features of Alberta oil sand bitumen and heavy oil asphaltenes. Energy Fuel 6(1):83–96
5. Wang Z, Liang W, Que G, Qian J (1997) Study on molecular structure of fractions in Shengli vacuum residue by ruthenium ions catalyzed oxidation. Acta Pet Sin (Pet Process Sect) 13(4):1–9
6. Peng PA, Fu J, Sheng G, Morales-Izquierdo A, Lown EM, Strausz OP (1999) Ruthenium-ions-catalyzed oxidation of an immature asphaltene: structural features and biomarker distribution. Energy Fuel 13(2):266–277
7. Strausz OP, Mojelsky TW, Faraji F, Lown EM, Peng PA (1999) Additional structural details on Athabasca asphaltene and their ramifications. Energy Fuel 13(2):207–227
8. Strausz OP, Mojelsky TW, Lown EM, Kowalewski I, Behar F (1999) Structural features of Boscan and Duri asphaltenes. Energy Fuel 13(2):228–247
9. Wang Z, Liang W, Que G, Qian J (1999) Investigation on chemical structure of fractions in Daqing vacuum residue by ruthenium ions catalyzed oxidation. J Fuel Chem Technol 27(2):102–109
10. Wang Z, Que G, Liang W, Qian J (1999) Investigation on chemical structure of resins and pentane asphaltenes in vacuum residua. Acta Pet Sin (Pet Process Sect) 15(6):39–46
11. Zhu J, Guo S, Li S (2002) Features of aromatic ring structure in petroleum asphaltene revealed by ruthenium Ion catalyzed oxidation. J Fuel Chem Technol 30(5):433–437
12. Djerassi C, Engle RR (1953) Oxidations with ruthenium tetroxide. J Am Chem Soc 75(15):3838–3840
13. Carlsen PHJ, Katsuki T, Martin VS, Sharpless KB (1981) A greatly improved procedure for ruthenium tetroxide catalyzed oxidations of organic compounds. J Org Chem 46(19):3936–3938
14. Stock LM, Tse K-T (1983) Ruthenium tetroxide catalysed oxidation of Illinois no. 6 coal and some representative hydrocarbons. Fuel 62(8):974–976
15. Mallya N, Zingaro RA (1984) Ruthenium tetroxide — a reagent with the potential for the study of oxygen functionalities in coal. Fuel 63(3):423–425
16. Stock LM, Wang S-H (1985) Ruthenium tetroxide catalysed oxidation of Illinois no. 6 coal: the formation of volatile monocarboxylic acids. Fuel 64(12):1713–1717
17. Stock LM, Wang S-H (1986) Ruthenium tetroxide catalysed oxidation of coals: the formation of aliphatic and benzene carboxylic acids. Fuel 65(11):1552–1562
18. Stock LM, Wang S-H (1987) The ruthenium(viii)-catalysed oxidation of Illinois no. 6 bituminous coal: an application of G.C.-Ft-I.R. spectroscopy for structural analysis. Fuel 66(7):921–924

19. Choi CY, Wang SH, Stock LM (1988) Ruthenium tetraoxide catalyzed oxidation of maceral groups. Energy Fuel 2(1):37–48
20. Stock LM, Wang SH (1989) Aliphatic structural elements of a Pocahontas no. 3 coal. Energy Fuel 3(4):533–535
21. Stock LM, Muntean JV (1993) Chemical constitution of Pocahontas no. 3 coal. Energy Fuel 7(6):704–709
22. Ilsley WH, Zingaro RA, Zoeller JH Jr (1986) The reactivity of ruthenium tetroxide towards aromatic and etheric functionalities in simple organic compounds. Fuel 65(9):1216–1220
23. Artok L, Murata S, Nomura M, Satoh T (1998) Reexamination of the Rico method. Energy Fuel 12(2):391–398
24. Murata S, U-Esaka K-I, Ino-Ue H, Nomura M (1994) Studies on aliphatic portion of coal organic materials based on ruthenium ion catalyzed oxidation. Energy Fuel 8(6):1379–1383
25. Nomura M, Artok L, Murata S, Yamamoto A, Hama H, Gao H, Kidena K (1998) Structural evaluation of Zao Zhuang coal. Energy Fuel 12(3):512–523
26. Nomura M, Kidena K, Hiro M, Murata S (2000) Mechanistic study on the plastic phenomena of coal. Energy Fuel 14(4):904–909
27. Kidena K, Bandoh N, Murata S, Nomura M (2001) Studies on the bond cleavage reactions of coal molecules and coal model compounds. Fuel Process Technol 74(2):93–105
28. Kidena K, Tani Y, Murata S, Nomura M (2004) Quantitative elucidation of bridge bonds and side chains in brown coals. Fuel 83(11–12):1697–1702
29. Su Y, Artok L, Murata S, Nomura M (1998) Structural analysis of the asphaltene fraction of an Arabian mixture by a ruthenium-ion-catalyzed oxidation reaction. Energy Fuel 12(6):1265–1271
30. Artok L, Su Y, Hirose Y, Hosokawa M, Murata S, Nomura M (1999) Structure and reactivity of petroleum-derived asphaltene. Energy Fuel 13(2):287–296
31. Murata S, Tani Y, Hiro M, Kidena K, Artok L, Nomura M, Miyake M (2001) Structural analysis of coal through Rico reaction: detailed analysis of heavy fractions. Fuel 80(14):2099–2109
32. Olson ES, Diehl JW, Froehlich ML, Miller DJ (1987) Elucidation of aliphatic structures in low-rank coals with ruthenium tetroxide oxidations. Fuel 66(7):968–972
33. Blanc P, Valisolalao J, Albrecht P, Kohut JP, Muller JF, Duchene JM (1991) Comparative geochemical study of three maceral groups from a high-volatile bituminous coal. Energy Fuel 5(6):875–884
34. Haenel MW (1992) Recent progress in coal structure research. Fuel 71(11):1211–1223
35. Standen G, Boucher RJ, Eglinton G, Hansen G, Eglinton TI, Larter SR (1992) Differentiation of German tertiary brown coal lithotypes ('Amorphous' and 'Woody' kerogens) using ruthenium tetroxide oxidation and pyrolysis-G.C.-M.S. Fuel 71(1):31–36
36. Shaohui G, Shuyuan L, Kuangzong Q (2001) Structural characterization of Chinese coal macerals by 13c Nmr and ruthenium ion catalyzed oxidation. Energy Sources 23(1):27–35
37. Petersen HI, Nytoft HP (2006) Oil generation capacity of coals as a function of coal age and aliphatic structure. Org Geochem 37(5):558–583
38. Akande SO, Ogunmoyero IB, Petersen HI, Nytoft HP (2007) Source rock evaluation of coals from the lower maastrichtian mamu formation, Se Nigeria. J Pet Geol 30(4):303–323
39. Huang Y-G, Zong Z-M, Yao Z-S, Zheng Y-X, Mou J, Liu G-F, Cao J-P, Ding M-H, Cai K-Y, Wang F, Zhao W, Xia Z-L, Wu L, Wei X-Y (2008) Ruthenium ion-catalyzed oxidation of Shenfu coal and its residues. Energy Fuel 22(3):1799–1806
40. Petersen HI, Lindstrom S, Nytoft HP, Rosenberg P (2009) Composition, peat-forming vegetation and kerogen paraffinicity of Cenozoic coals: relationship to variations in the petroleum generation potential (hydrogen index). Int J Coal Geol 78(2):119–134
41. Yao Z-S, Wei X-Y, Huang Y-G, Zong Z-M, Huang Y, Xu J-J, Li Y, Lu Y, Lv J, Lu H (2009) Compositional and structural features of the extracts from Shenfu coal. J Wuhan Univ Sci Technol 32(6):631–637

42. Yao Z-S, Wei X-Y, Lv J, Liu F-J, Huang Y-G, Xu J-J, Chen F-J, Huang Y, Li Y, Lu Y, Zong Z-M (2010) Oxidation of Shenfu coal with Ruo4 and naocl. Energy Fuel 24:1801–1808
43. Ma L, Lu D-R, Li S, Liang H-D, Zhu S-Q (2013) Ft-Icr Ms analytical study on the products of Shenhua coal using ruthenium-ion-catalyzed oxidation method. J China Coal Soc 38 (S1):223–230
44. Ma L, Lu D-R, Liang H-D, Zhu S-Q, Ding Y, Li S, Chen Y-F (2013) Preliminary study on macromolecular structure characteristics of Shenhua long flame coal. J Fuel Chem Technol 41(5):513–522
45. Muhammad AB, Abbott GD (2013) The thermal evolution of asphaltene-bound biomarkers from coals of different rank: a potential information resource during coal biodegradation. Int J Coal Geol 107:90–95
46. Payzant J, Lown E, Strausz O (1991) Structural units of Athabasca asphaltene: the aromatics with a linear carbon framework. Energy Fuel 5(3):445–453
47. Strausz OP, Mojelsky TW, Lown EM (1992) The molecular structure of asphaltene: an unfolding story. Fuel 71(12):1355–1363
48. Peng PA, Morales-Izquierdo A, Hogg A, Strausz OP (1997) Molecular structure of Athabasca asphaltene: sulfide, ether, and ester linkages. Energy Fuel 11(6):1171–1187
49. Mullins OC (2007) Rebuttal to comment by professors Herod, Kandiyoti, and Bartle on "molecular size and weight of asphaltene and asphaltene solubility fractions from coals, crude oils and bitumen". Fuel 86(1–2):309–312
50. Mullins OC, Martínez-Haya B, Marshall AG (2008) Contrasting perspective on asphaltene molecular weight. This comment vs the overview of A. A. Herod, K. D. Bartle, and R. Kandiyoti. Energy Fuel 22(3):1765–1773
51. Mullins OC (2009) Rebuttal to Strausz et al. regarding time-resolved fluorescence depolarization of asphaltenes. Energy Fuel 23(5):2845–2854
52. Mullins OC (2010) The modified Yen model. Energy Fuel 24(4):2179–2207
53. Herod AA, Bartle KD, Kandiyoti R (2007) Characterization of heavy hydrocarbons by chromatographic and mass spectrometric methods: an overview. Energy Fuel 21(4):2176–2203
54. Herod AA, Bartle KD, Kandiyoti R (2008) Comment on a paper by Mullins, Martinez-haya, and Marshall "contrasting perspective on asphaltene molecular weight. This comment vs the overview of A. A. Herod, K. D. Bartle, and R. Kandiyoti". Energy Fuel 22(6):4312–4317
55. Herod AA, Kandiyoti R (2008) Comment on "Limitations of Size-Exclusion Chromatography in Analyzing Petroleum Asphaltenes: A Proof by Atomic Force Microscopy" by Behrouzi M Luckham PF. Energy Fuels 22(3):1792–1798. doi:10.1021/Ef800064q Energy Fuels 2008;22(6):4307–4309
56. Strausz OP, Safarik I, Lown EM, Morales-Izquierdo A (2008) A critique of asphaltene fluorescence decay and depolarization-based claims about molecular weight and molecular architecture. Energy Fuel 22(2):1156–1166
57. Strausz OP, Safarik I, Lown EM (2009) Cause of asphaltene fluorescence intensity variation with molecular weight and its ramifications for laser ionization mass spectrometry. Energy Fuel 23(3):1555–1562
58. Zhu J, Li S, Guo S (2003) New methods for the study of biodegraded crude oil. J Fuel Chem Technol 31(01):1–5
59. Ma A, Zhang S, Zhang D (2004) Ruthenium-Inos-catalyzed oxidation of the asphaltenes of heavy oils from Lunnan and Tahe oil fields of the Tarim Basin NW China. Nat Gas Geosci 15 (02):144–149
60. Ma A, Zhang S, Zhang D, Lu G (2004) Ruthenium-ions-catalyzed oxidation of the asphaltenes of oils and oil-source correlation in the Tarim Basin. Pet Explor Dev 31 (03):54–58
61. Ma A, Zhang S, Zhang D, Jin Z (2005) The advances in the geochemistry of the biodegraded oil. Adv Earth Sci 20(04):449–454

62. Ma A, Zhang S, Zhang D, Jin Z, Chen Z (2005) Ruthenium-ions-catalyzed oxidation of an asphaltene of a biodegraded oil from Caoqiao oilfield, Dongying depression, Bohaiwan basin-the distribution of biomarkers and the geological significance. Pet Geol Exp 27(3):288–292
63. Xiong Y, Wang Y, Wang Y (2007) Selective chemical degradation of kerogen from nenjiang formation of the Southern Songliao basin. Sci China Ser D Earth Sci 50(10):1504–1509
64. Barakat AO, Scholz-Böttcher BM, Rullkötter J (2013) Structural investigations of Monterey kerogen by sequential chemical degradation. Fuel 104:788–797
65. Liu F-J, Wei X-Y, Gui J, Wang Y-G, Li P, Zong Z-M (2013) Characterization of biomarkers and structural features of condensed aromatics in Xianfeng lignite. Energy Fuel 27(12):7369–7378
66. Cyr N, Mcintyre D, Toth G, Strausz O (1987) Hydrocarbon structural group analysis of Athabasca asphaltene and its Gpc fractions by 13c Nmr. Fuel 66(12):1709–1714
67. Schuda PF, Cichowicz MB, Heimann MR (1983) A facile method for the oxidative removal of benzyl ethers: the oxidation of benzyl ethers to benzoates by ruthenium tetraoxide. Tetrahedron Lett 24(36):3829–3830
68. Zhou X, Shi Q, Zhang Y, Zhao S, Zhang R, Chung KH, Xu C (2012) Analysis of saturated hydrocarbons by redox reaction with negative-ion electrospray fourier transform ion cyclotron resonance mass spectrometry. Anal Chem 84(7):3192–3199
69. Bakke JM, Frøhaug AE (1996) Ruthenium tetraoxide mediated reactions: the mechanisms of oxidations of hydrocarbons and ethers. J Phys Org Chem 9(6):310–318
70. Bakke JM, Frøhaug AE (1996) Mechanism of Ruo4-mediated oxidations of saturated hydrocarbons, isotope effects, solvent effects and substituent effects. J Phys Org Chem 9(7):507–513
71. Dragojlović V, Bajc S, Amblès A, Vitorović D (2005) Ether and ester moieties in Messel shale kerogen examined by hydrolysis/ruthenium tetroxide oxidation/hydrolysis. Org Geochem 36(1):1–12
72. Bakke JM, Lundquist M (1986) The Ruo4 oxidation of cyclic saturated hydrocarbons formation of alcohols. Acta Chem Scand 40B:430–433
73. Zhou X, Zhang Y, Zhao S, Chung KH, Xu C, Shi Q (2013) Characterization of saturated hydrocarbons in vacuum petroleum residua: redox derivatization followed by negative-Ion electrospray ionization Fourier transform ion cyclotron resonance mass spectrometry. Energy Fuel 28(1):417–422
74. Strausz OP, Lown EM (2003) The chemistry of Alberta oil sands, bitumens and heavy oils. Alberta Energy Research Institute, Calgary
75. Zhou X (2013) Characterization of molecular structure in heavy oil by ruthenium ion catalyzed oxidation, in College of Chemical Engineering, China University of Petroleum, Beijing
76. Mojelsky TW, Montgomery DSS, Otto P (1985) Ruthenium (VIII) catalyzed oxidation of high molecular weight components of Athabasca Oil Sand Bitumen. Aostra J Res 2(2):131–137
77. Zijun W, Wenjie L, Guohe Q, Jialin Q (1997) Structural characterization of gudao asphaltene by ruthenium ion catalyzed oxidation. Pet Sci Technol 15(5-6):559–577
78. Zhang ZG, Guo S, Zhao S, Mou T (2006) Structure features of the supercritical fluid extraction and fraction tailing of Dagang vacuum residue. J Fuel Chem Technol 34(4):427–433
79. Zhang H, Yan Y, Cheng Z, Sun W, Guan M (2007) Changes of asphaltene after hydrotreating by ruthenium ions catalyzed oxidation. Acta Pet Sin (Pet Process Sect) 23(4):33–38
80. Zhang ZG, Guo S, Yan G, Zhao S, Song L, Chen L (2007) Distribution of polymethylene bridges and alkyl side chains in dagang vacuum residue asphaltene and Sfef tailing asphaltene. J Chem Ind Eng (China) 58(10):2601–2607
81. Zhang ZG, Guo S, Yan G, Zhao S, Song L, Chen L (2007) Chemical structural features of fractions from dagang vacuum residue. J Fuel Chem Technol 35(5):553–557

82. Zhang ZG, Guo S, Zhao S, Yan G (2007) Chemical structure features of polar fractions of Sfef tailing from dagang vacuum residue. Acta Pet Sin (Pet Process Sect) 23(4):82–88
83. Zhang H, Yan Y, Cheng Z, Sun W (2008) Structural changes of sub-fractions in residue hydrotreating products by ruthenium catalyzed oxidation. Pet Sci Technol 26(16):1945–1962
84. Ma A, Zhang S, Zhang D, Jin Z (2004) Oil and source correlation in Lunnan and Tahe heavy oil fields. Oil Gas Geol 25(1):31–38
85. Ali MF, Siddiqui MN, Al-Hajji AA (2004) Structural studies on residual fuel oil asphaltenes by Rico method. Pet Sci Technol 22(5-6):631–645
86. Jia W, Peng PA (2004) Molecular structure of oil asphaltenes from Lunnan area of the Tarim Basin and its applications: a study by pyrolysis, methylation-pyrolysis and Rico. Geochim 33(2):139–146
87. Ma A, Zhang S, Zhang D (2008) Ruthenium-ion-catalyzed oxidation of asphaltenes of heavy oils in Lunnan and Tahe oilfields in Tarim Basin NW China. Org Geochem 39(11):1502–1511
88. Zhang H, Yan Y, Cheng Z, Sun W (2009) Structural analysis of coke on used catalysts during residue hydrotreating by ruthenium ion catalyzed oxidation reaction. Pet Sci Technol 27(1):33–45
89. Barakat AO, Scholz-Boettcher BM, Rullkoetter J (2012) Ruthenium tetroxide oxidation of immature sulfur-rich kerogens from the Nordlinger Ries (Southern Germany). Fuel 96(1):176–184
90. Khaddor M, Ziyad M, Amblès A (2008) Structural characterization of the kerogen from youssoufia phosphate formation using mild potassium permanganate oxidation. Org Geochem 39(6):730–740
91. Yoshioka H, Ishiwatari R (2005) An improved ruthenium tetroxide oxidation of marine and lacustrine kerogens: possible origin of low molecular weight acids and benzenecarboxylic acids. Org Geochem 36(1):83–94
92. Li C, Peng P, Sheng GY, Fu JM (2004) A study of a 1.2 Ga kerogen using Ru ion-catalyzed and pyrolysis-gas chromatography-mass spectrometry: structural features and possible source. Org Geochem 35(5):531–541
93. Kribii A, Lemee L, Chaouch A, Ambles A (2001) Structural study of the Moroccan timahdit (Y-layer) oil shale kerogen using chemical degradations. Fuel 80(5):681–691
94. Reiss C, Blanc P, Trendel JM, Albrecht P (1997) Novel hopanoid derivatives released by oxidation of Messel shale kerogen. Tetrahedron 53(16):5767–5774
95. Boucher RJ, Standen G, Eglinton G (1991) Molecular characterization of kerogens by mild selective chemical degradation — ruthenium tetroxide oxidation. Fuel 70(6):695–702
96. Boucher RJ, Standen G, Patience RL, Eglinton G (1990) Molecular characterisation of kerogen from the kimmeridge clay formation by mild selective chemical degradation and solid state 13c-Nmr. Org Geochem 16(4–6):951–958
97. Guo S, Li S, Qin K (2000) Structural characterization of kerogen and macerals by ruthenium ion catalyzed oxidation. J Univ Pet China 24(3):54–57
98. Standen G, Boucher RJ, Rafalska-Bloch J, Eglinton G (1991) Ruthenium tetroxide oxidation of natural organic macromolecules: messel kerogen. Chem Geol 91(4):297–313
99. Blokker P, Van Bergen P, Pancost R, Collinson ME, De Leeuw JW, Damste JSS (2001) The chemical structure of Gloeocapsomorpha Prisca microfossils: implications for their origin. Geochim Et Cosmochim Acta 65(6):885–900
100. Marshall AG, Rodgers RP (2004) Petroleomics: the next grand challenge for chemical analysis. Acc Chem Res 37(1):53–59
101. Rodgers RP, Schaub TM, Marshall AG (2005) Petroleomics: Ms returns to its roots. Anal Chem 77(1):20 A–27 A
102. Shi Q, Zhang Y, Xu C, Zhao S, Chung KH (2014) Progress and prospect on petroleum analysis by Fourier transform ion cyclotron resonance mass spectrometry. Sci China Chem (in Chinese) 44(5):694–700

Molecular-Level Composition and Reaction Modeling for Heavy Petroleum Complex System

Zhen Hou, Linzhou Zhang, Scott R. Horton, Quan Shi, Suoqi Zhao, Chunming Xu, and Michael T. Klein

Abstract A new methodology for the molecule-based modeling of heavy petroleum mixtures has been developed. Molecules in the heavy feedstock have been described in terms of three essential structural attributes (cores, side chains, and inter-core linkages) and then statistically juxtaposed into a set of representative molecular compositions that can be constrained by a set of probability density functions (pdfs). In order to obtain the optimal molecular composition, an optimization loop was employed to minimize an objective function in terms of available measurements via adjusting the limited parameters of the pdfs. An example of resid feedstock containing 400,000 components was created using only O(30) parameters. To limit the kinetic model to a practical size, the reaction model was described in terms the reactions of the three constituent types and a set of irreducible molecules. Subsequent product property estimation was a straightforward juxtaposition of attributes. For example, a model for resid coking was built in terms of 2,839 attributes and equations but kept the full compositional details of the 400,000-molecule mixture.

Keywords Composition modeling · Heavy oil · Kinetic modeling · Resid

---

Z. Hou, S.R. Horton, and M.T. Klein (✉)
Department of Chemical and Biomolecular Engineering, University of Delaware Energy Institute, University of Delaware, Newark, DE 19716, USA
e-mail: houzhen@udel.edu; scottrhorton@gmail.com

L. Zhang, Q. Shi, S. Zhao, and C. Xu
State Key Laboratory of Heavy Oil Processing, China University of Petroleum, Beijing 102249, China
e-mail: linzhou.zhang.china@gmail.com; sq@cup.edu.cn; zhaosuoqi@vip.sina.com; xcm@cup.edu.cn

Contents

1 Introduction 94
2 Composition Model for Heavy Oil 96
   2.1 Qualitative Molecular Information Determination 96
   2.2 Quantitative Molecular Information Determination 104
3 Reaction Model for Heavy Oil 112
   3.1 ARM Reaction Network Analysis 112
   3.2 ARM Model Equation 114
   3.3 Kinetic Parameters: LFER 115
   3.4 Post-reaction Sampling and Product Property Estimation 116
   3.5 Representative Results of a Resid Pyrolysis Model 117
4 Summary 117
References 118

1 Introduction

The creation of energy solutions is one of humanity's top ten problems for the next 50 years and is being studied worldwide. Although developing alternative resources (coal, biomass) is a promising solution, the deep utilization of heavy petroleum is one of the most inexpensive and immediate ways to increase energy supplies. The development of detailed kinetic and process models will help to optimize the heavy oil conversion: a detailed model of heavy oil conversion could not only support routine engineering goals (the prediction of the product properties, yields, operation condition, etc.) but also provide the essential understanding of its chemistries and thus help to improve catalyst development, process design, and optimization.

Molecular-level modeling is an optimal starting point. A collection of molecules provides the essential structural information in a complex hydrocarbon mixture and serves as a basis for property estimation, thermodynamics, and chemical kinetics. As a result, a molecule-based model can link the various levels of chemical study, from quantum chemical calculations to processing chemistries.

In the area of molecular-level modeling for complex process chemistries, several research groups have developed different approaches. Here we will consider contributions from Froment and Klein's groups and ExxonMobil, IFP, and UMIST laboratories.

Froment [1–3] and his coworkers developed a single-event approach that uses graph theory and can build fundamental kinetic models at the mechanistic level. This approach decomposes every pathway level reaction into a set of elementary mechanistic steps. Based on the intrinsic entropy and the reaction coordinate of each elementary step, a set of elementary mechanism steps and derived single-event frequency factors (A) were obtained. The frequency factor of an elementary step was expressed as a function of the change of intrinsic entropy that was estimated by quantum chemistry packages (e.g., GAUSS, MOPAC). The activation energy was calculated via Evans–Polanyi correlations [4]. This approach is ideal for light end fraction conversion but can be difficult to extend to heavy-end conversion because of the size of the modeling problem. The elemental mechanism step analysis of the reaction path and its related quantum chemistry calculations for heavy oil molecules still represent a challenging computational burden even for state-of-the-art computer hardware.

Quann and Jaffe [5, 6] of ExxonMobil developed the structure-oriented lumping (SOL) approach that uses a set of structural vectors to describe petroleum components at the molecular level as well as the kinetics models for petroleum process chemistries. SOL provides a leading example from industry to develop and apply practical molecule-based models to petroleum hydrocarbon conversions. SOL was successfully applied to middle to VGO fractions. An extension to SOL [7] was developed to describe heavy resid.

Towler and colleagues at UMIST (University of Manchester Institute of Science and Technology) developed a MTHS [8] (molecular-type homologous series) matrix to represent the species and built kinetics model based on this MTHS representation. The components on MTHS are in a closed matrix and thus it is computationally tractable. This approach is applicable for gasoline and diesel [9–12] but it is hard to extend to heavy resid.

Hudebine et al. from IFP [13–17] used a stochastic approach to represent molecular components in a feedstock and applied Froment's approach to the kinetics model. This approach was applicable to gas oil, VGO, and vacuum resid (VR).

The Klein Research Group (KRG) represents hydrocarbon molecules by bond-electron (BE) matrices. Any chemical reactions, namely, the bond-making and bond-breaking processes, are also implemented as reaction matrix operations. A reactant to product relationship can be derived by performing a matrix addition of the reaction matrix to the reactant matrix in order to obtain a product species matrix. This approach has been developed as an automated reaction network generation tool called NetGen [18–20] and later upgraded to an enhanced version InGen [21] that enables the description of most complex chemistries for petroleum hydrocarbon conversions [22–27]. At the early stage, KRG used a stochastic approach (MolGen) [28, 29] to obtain the initial conditions of kinetic modeling: the molecular compositions of a feedstock. Recently, this approach has been replaced by a new composition model tool called CME [30]. CME describes a complex feedstock as a set of homologous series constrained by a set of statistical probability density functions (pdfs) with limited parameters. CME derives the optimal molecular compositions of a feedstock as BE matrices that can be incorporated into kinetic modeling seamlessly. Pyl and Zhen [30, 31] applied this approach to kerosene and heavy gas oil fractions.

Although the above research groups and others have developed and applied molecule-based modeling technology in petroleum hydrocarbon mixtures for the past decades, the extension of the molecular-level modeling to heavy oils is still a challenging problem arising from the staggering complexity of not only the reaction mixtures but the complexity of each molecule within the mixture.

Heavy complex feeds often have low volatility with high boiling points, and thus complete sampling using the current analytical chemistry techniques is not possible. Therefore, it is hard to obtain a full set of molecular representations by direct measurements for such complex mixtures. In addition, heavy resids often contained thousands of "multifunctional" large molecular components such as those illustrated in Fig. 1. Analyzing chemical functions and structure characteristics of such large complex molecules digitally will require large computation. Generally, a heavy complex

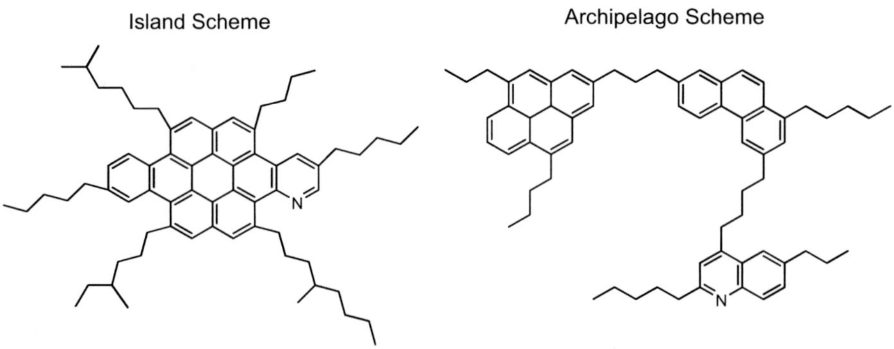

Fig. 1 Typical molecular structures in heavy resid

mixture like petroleum resid will contain 50,000 or more such molecular species, which is far beyond current computational tractability. Consequently, the large number of species in heavy resid is an impractical computational issue for reaction and kinetic models. Traditional deterministic kinetic models consist of one ordinary differential equation (ODE) per each species, and thus the numerical burden of solving 50,000 simultaneous ODEs is beyond the upper limit of what is considered practical. Therefore, the sheer size of the thus-implied modeling problem engenders a conflict between the need for molecular detail and the formulation and solution of the model.

In order to address this dilemma, KRG developed a new approach to model a heavy petroleum mixture at the molecular level with a practical computational demand (e.g., 5,000 species [6]). Part of this work was summarized in recent publications [30–34]. In this article, we systematically illustrate our current modeling methodology in both compositional modeling and kinetic modeling for heavy oil conversion.

2 Composition Model for Heavy Oil

As mentioned above, a full detailed molecular representation of heavy oil from direct measurements is not yet practical. Therefore, we need to develop a logical framework to convert indirect measurements, such as those shown in Table 1, into an optimal molecular composition of a heavy resid. The approach we have developed is classified in two steps: qualitative molecular information (molecular structures) determination and quantitative molecular information (mole or weight fractions).

2.1 Qualitative Molecular Information Determination

In order to represent thousands of large multiple components of heavy oils, the molecular structures of resid components were abstracted into three basic structural attributes: cores, side chains (SCs), and inter-core linkages (IL). Cores represent

Table 1 Typical measurements for heavy resid

#	Measurement name	Obtained property name	Comments
1	Density	Density	Standard method, good repeatability, high accuracy
2	Elemental analysis	C, H, S, N, O elemental composition	Standard method; for C, H, S, and N good repeatability; O a little bit low repeatability
3	SARA	Saturates, aromatics, resins, asphaltene composition	Standard method, open column chromatography, low repeatability
4	VPO	Mn	Standard method, moderate repeatability, low accuracy
5a	GPC	Molecular weight distribution	Standard method, good repeatability, moderate accuracy
5b	Low-resolution MS (ToF)	Molecular weight distribution	Standard method, moderate repeatability, moderate accuracy
6	SimDis	Boiling point distribution	Using GC, good repeatability
7	High-temperature SimDis	Boiling point distribution	Using GC, good repeatability, not a common instrument in general lab
8	Acid–base titration	Content of basic nitrogen	Standard method, low repeatability, low accuracy
9	NMR	Fractions of aromatic, naphthenic, paraffinic carbon	For 1H-NMR, high accuracy with semiquantitative information; for 13C-NMR, moderate accuracy with quantitative information
10	HPLC	Aromatic ring type analysis (1 ring, 2 ring, 3 ring, etc.)	
11	RICO	Side-chain distribution/linkage chain distribution	Time-consuming reaction/separation procedure, moderate repeatability
12	XPS	Functional group	The data processing is complex and needs advanced knowledge
13	High-resolution MS (FTICR-MS)	Semiquantitative detailed molecular composition	

ring and aggregated fused ring structures. SC represents the free-terminal substituents attached to cores, whereas IL represents the substituents between two cores. An associated abstraction of a resid molecule is shown in Fig. 2. We will illustrate how to determine core, SC, and IL, respectively, and then show our approach to represent a resid molecule via those three attributes.

2.1.1 Core Representation

Core is the most important and complex structural attribute in the heavy oil system. Their large fused ring structures play a critical role in influencing the kinetic reactivity, thermodynamics, and physical properties of a heavy molecule. Many

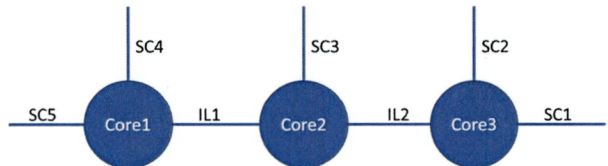

Fig. 2 Abstraction of molecular structures in heavy oil

researchers often used the type of core (called molecular type) to classify and describe molecular compositions into a set of homologous series [35–37].

In recent years, the successful application of advanced analytical techniques such as Fourier transform ion cyclotron resonance mass spectrometry (FTICR-MS) has provided clear recognition of the structural features of heavy oil in terms of carbon number (or molecular weight), DBE (or Z number), chemical formula, and compound class [38–48]. Particularly, collision-induced dissociation (CID) and infrared multiphoton dissociation (IRMPD) [41, 44, 49] can specifically provide structural information for dealkylated core structures.

However, this structural information cannot directly provide information on core structures in heavy oil. The number of isomeric ring structures grows dramatically with the ring size. So it is impractical to represent all molecular structures for large fused ring structures in heavy oil. In addition, an atomically explicit representation (e.g., BE matrix) of a large ring structure requires complicated computation and is thus hard to manipulate explicitly. A set of structural building blocks provides a concise and efficient way to represent core structures. ExxonMobil SOL vectors [5, 6] provide a leading example of using building blocks to describe hydrocarbon mixtures. Based on similar structural features, we developed an application called CoreGen [30] that provides a set of irreducible hydrocarbon building blocks (shown in Fig. 3) to solely represent core structures in complex petroleum system. The building rules are based on previous kinetic and thermodynamic expertise and also analytical findings in petroleum mixtures. As a result, an optimal molecular structure of cores was obtained from one unique building block configuration.

Figure 4 illustrates how to use CoreGen and its irreducible hydrocarbon building blocks to determine core structures of a complex heavy oil feedstock.

From detailed advanced measurements such as CID-FTICR-MS, we can qualitatively acquire the detailed structural information including the range of DBE, carbon number, and hetero compound class and thus derive the limits of core structure such as aromatic ring size, naphthenic ring size, heteroatom ring size, etc. As a result, we can set a limited numerical value range for each hydrocarbon building block. Within

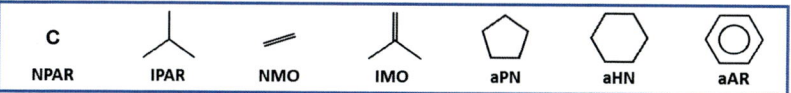

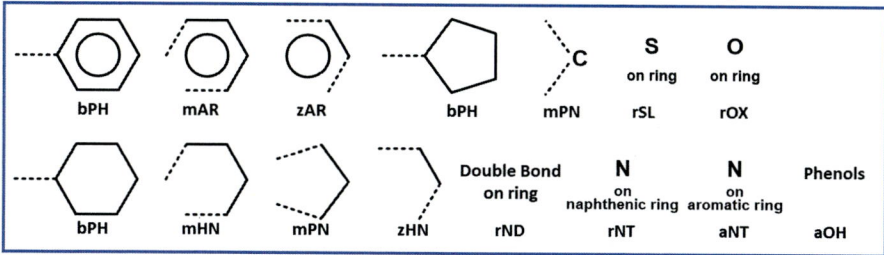

Fig. 3 Irreducible hydrocarbon building blocks in CoreGen

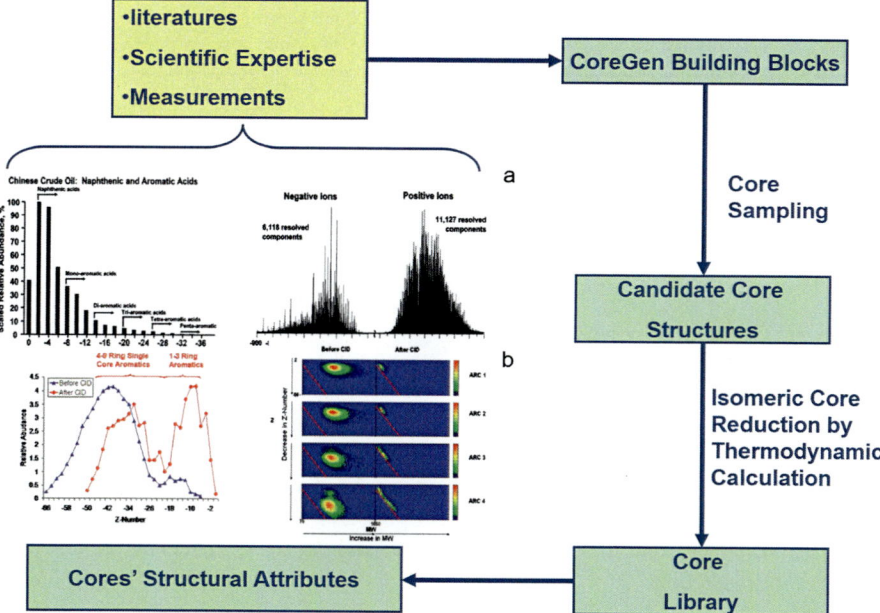

Fig. 4 Flowsheet of core determination for heavy oil feedstocks. (**a**) See Marshall and Rodgers [50]. (**b**) See Qian et al. [49]

those numerical limitations, we can sample a list of possible candidate core structures by traversing the combination of different building block configurations. The statistics of sampling candidate core structures is shown in Fig. 5.

In this sampling example, we set the following limits: the size of the aromatic ring < 9, the size of the naphthenic ring (six-membered ring only) < 6, the size of

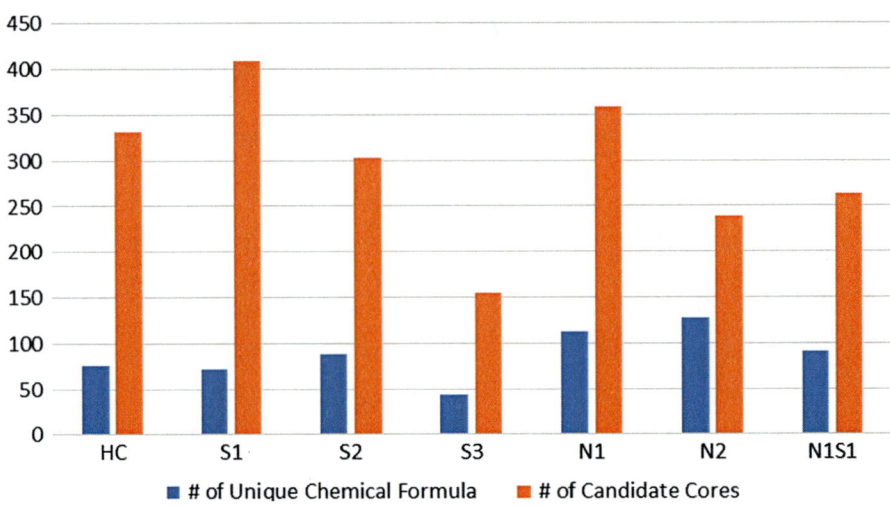

Fig. 5 A statistics of candidate cores via core building block traverse

Table 2 Selected isomeric core reduction criteria: kinetics and thermodynamics

Name of selected criteria	Description
Thermodynamic stability	Gibbs formation, heat of formation, entropy
LFER reactivity	Highest bond order, PI electron density, heat of formation, sulfur electron density, etc.
Bulk properties	Boiling point, density
Visualized structures	Experienced structure pattern found in crude oils

the thiophenic ring < 3, and the size of the nitrogen-containing ring < 2. The number of candidate cores is O(1000) but the number of unique chemical formulae corresponding to those candidate cores is limited to O(100). If we perform a further isomeric core reduction to assign one core to one chemical formula derived from the significant data points of high-resolution MS (e.g., FTICR-MS) results, we could obtain a practical size of cores, which may be used for further composition and kinetic modeling. The criteria of isomeric core reduction is determined by kinetic reactivity, thermodynamics, physical properties, and experienced structural patterns from previous crude oil analyses in the past decades. CoreGen can directly convert candidate core structures to an atom-explicit representation (BE matrix) and thus provides a straightforward way to evaluate thermodynamics, physical properties, and kinetic reactivity via atomic-based fragmental group contribution methods (e.g., Benson method [51], Gani method [52]) and quantum chemical calculations (e.g., MOPAC http://openbabel.org/wiki/Main_Page). A selected criterion to perform isomeric core reduction is shown in Table 2.

After isomeric core reduction, a core structure library was constructed to represent optimal core structures in a heavy feedstock. A typical core library of a resid [32] is shown in Figs. 6 and 7.

Fig. 6 Selected hydrocarbon cores in core library. Reprinted with permission from [32]. Copyright (2014) American Chemical Society

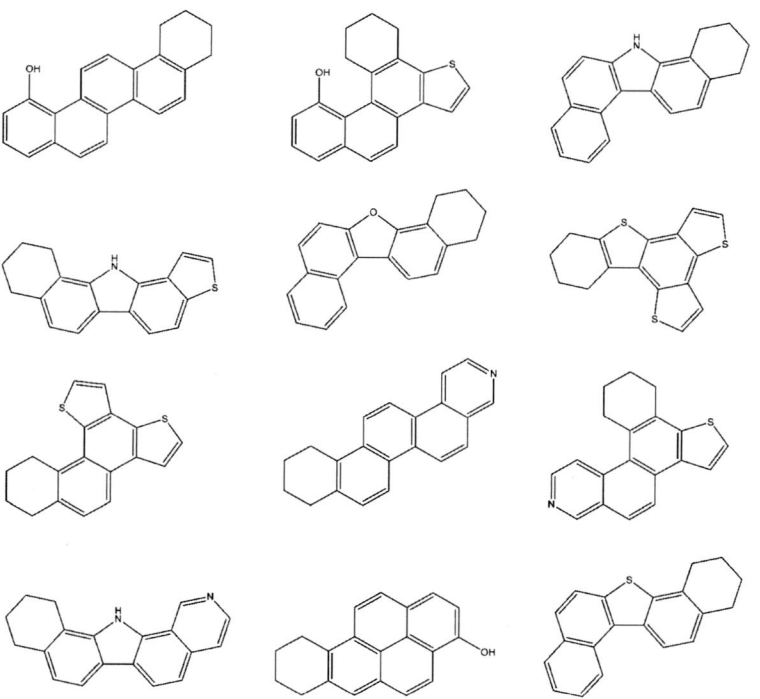

Fig. 7 Selected heteroatom cores in core library. Reprinted with permission from [32]. Copyright (2014) American Chemical Society

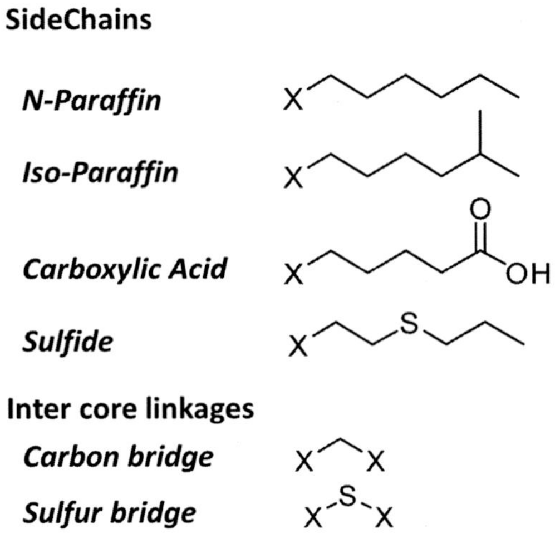

Fig. 8 Selected SCs and ILs in SC and IL library. Reprinted with permission from [32]. Copyright (2014) American Chemical Society

SideChains

N-Paraffin

Iso-Paraffin

Carboxylic Acid

Sulfide

Inter core linkages

Carbon bridge

Sulfur bridge

In this example, we constructed 373 cores including pure hydrocarbon ring systems and sulfur-, nitrogen-, and oxygen-containing ring structures. The numbers of aromatic rings and naphthenic rings were limit to nine and six, respectively. Only six-membered naphthenic ring structures were considered. Both thiophenic and cyclo-sulfide rings were considered as sulfur-containing cores, and the sulfur species were limited to S1, S2, and S3 classes. The nitrogen-containing cores were limited to N1 and N2 classes and considered as pyrrolic and pyridinic structures. Oxygen-containing cores were considered as phenol and furan structures and limited to O1 classes. The multi-heteroatom cores such as N1S1 were also included in the library.

Compared to core determination, the identification of SC and IL is a relatively straightforward process.

2.1.2 SC and IL Representation

The SCs of resid components were pre-classified as a set of discrete types. In each type, the length of SC was assumed to grow by CH2 groups. The IL of resid structures is similar for SC but does not have free terminal groups (e.g., CH_3) at the end.

A typical example [32] of SC and IL for heavy resid is shown in Fig. 8. In this resid example, four types of SC were selected based on current FTICR-MS results: n-paraffin, isoparaffin, sulfide, and carboxylic acid. In order to simplify the model, we assumed branched methyl and COOH groups were linked at the end of SC and assigned sulfide groups into the middle of SC. The length of SC was limited to 50. IL structures were limited to one carbon bridge and one sulfur bridge [53–55].

A summary of all cores, SCs, and ILs in an example resid was shown in Table 3.

Table 3 A summary of cores, SCs, and ILs in a resid example model [32]

Attribute type	Category	# of attribute	Description
Cores	Pure hydrocarbon cores	44	Max naphthenic ring # is 6 Max aromatic ring # is 9
	Sulfur rings	89	# of S atom from 1–3
	Nitrogen rings	86	# of N atom from 1–3
	Oxygen rings	68	Furan and phenol
	Multi-heteroatom rings	86	For example, S1N1, etc.
Total # of cores		373	
SC	n-Paraffin	50	C# range is 1–50
	*Iso*paraffin	47	C# range is 4–50, one branched methyl at the end of the chain
	Sulfide	47	C# range is 4–50, one sulfide at the middle of the chain
	Carboxylic acid	47	C# range is 4–50, one COOH group at the end of the chain
Total # of SCs		191	
IL		2	C1 and S bridge

With the cores, SCs, and ILs of a resid feedstock now represented, the next step is to use this information to construct molecular components for a resid feedstock.

2.1.3 Resid Molecule Sampling

Although the state-of-the-art analytical techniques gave us more clear information about the structural attributes (e.g., core, SC, and IL), it is still unable to measure the differences of the substituted position at which core, SC, and IL attached. In addition, those isomeric substitution patterns do not play a major role in the process of resid upgrading. We can thus ignore such substituent patterns, which leads to significant simplifications in the resid model.

Island and archipelago models are two structural schemes to describe the resid molecules. In general, a model should handle both of these schemes and limiting cases where one or the other predominates. Thus, the framework of the archipelago scheme is the optimal starting point for a general approach because the island model is a special case when the degree of polymerization (the cluster size) is limited to one. Referring to linear polymer theory, we employed a Bethe linear polymer [56] scheme to describe the large molecule in resid system. As shown in Fig. 2, a generic heavy resid molecule can be treated as a linear polymer. Each core in a molecule can be regarded as a monomer. As a normal linear polymer, there is only one IL connected between two cores. Given a certain cluster size n, there are n cores and (n−1) ILs in a heavy resid molecule. Following the polymer statistical modeling, we can set a binding site number (coordination number) on a core, which indicates the number of substituents (SCs or ILs) attached to a core. Given a certain coordination number m, the terminal core will attach (m−1) SCs on it and other intermediate cores will attach (m−2) SCs on it. Ignoring the connection details among cores, SCs, and ILs, we can sample resid molecular components as a combination of cores, SCs, and ILs following the above sampling rules.

This sampling protocol will sample cores and ILs to a set of non-SC core clusters first and attach SCs later. As an example, we sampled around 2,260 non-SC core clusters in the resid model [32] discussed above when limiting the maximum cluster size to 3. If we treated those over 2,000 non-SC core clusters as "molecular type," we will obtain an extended homologous series with those non-SC core clusters varying the carbon number increments of SCs.

To perform a fully statistical SC (carbon number limit is 50) sampling will cause an extremely large combinatorial computation and thus far exceed our current computer hardware capacity. To address this issue, we simplified SC sampling with a new algorithm called main methyl method (MMM) [33]. In a generic resid molecule, MMM assumes that there is only one main SC that could be any length attached to a core and other binding sites of core are connected with a methyl group only. MMM reduces the combinatorial calculation of the SC sampling and provides us a practical computational burden. As a result, we finally sample out around 400,000 representative molecular components in an example resid [32].

Because the example resid [32] provides very detailed measurement data with a very high boiling range sample (up to 700C), we performed a very detailed sampling and used 400,000 components to match the simulated distillation curve and show the detailed molecular weight and DBE distribution. In practical cases, the sampling size can be reduced by limiting a set of physical conditional constraints (e.g., boiling point, DBE, aromatic ring size, molecular weight, etc.) while keeping the same sampling rule.

The next step is to determine the mole fraction/weight fraction of molecular components in a feedstock to finish the compositional model.

2.2 Quantitative Molecular Information Determination

2.2.1 Quantitative Sampling Protocol

In order to describe and constrain the large number of resid molecular components, a statistical mathematical method was employed. We can use one or a set of probability density functions (pdfs) to describe cores, SCs and ILs, respectively, and apply a statistical algorithm to calculate all individual molecular components derived from them.

The types of SCs and ILs can be treated as a histogram pdf with a set of discrete values. This is because the number of types is small ($<=4$). Much of the relevant literature [28–31] shows the length of SCs and ILs can be constrained by a gamma pdf, and this notion is also supported by experimental data [32, 35].

The pdfs of SCs and ILs for an example resid model [32] are shown in Table 4. In order to simplify the model, we assumed the length of IL pdf and the types of IL

Table 4 Selected attribute pdfs for an example resid

Attribute type	Constrained attributes	PDF type	# of parameter	Description
Core	Core type	Histogram	2	Three types: paraffin, naphthenes, and aromatics. # of parameter is the degree of freedom 2 (DOF): 3−1=2
	Naphthenic ring #	Gamma	2	Min. in gamma function is determined by physical ring size and thus only two parameters for this gamma function: mean and stdev
	Aromatic ring #	Gamma	2	Min. in gamma function is determined by physical ring size and thus only two parameters for this gamma function: mean and stdev
	Sulfide ring #	Conditional	2	
	Thiophene ring #	Conditional	2	
	Pyridine ring #	Conditional	2	
	Pyrrole ring #	Conditional	2	
	Phenol group	Conditional	2	
	Ring configuration	Gamma	2	By default, the gamma distribution of C# is applied. Min. in gamma function is determined by physical ring size and thus only two parameters for this gamma function: mean and stdev
# of parameters for core			18	
SC	Side-chain type and length of chain	Histogram	3	4 SC types and the number of parameters is the DOF: 4−1=3
	Length of chain	Gamma	2	Min. in gamma function is determined by physical SC size and thus only two parameters for this gamma function: mean and stdev
# of parameters for SC			5	
IL	Inter-core linkage type	Histogram	1	

(continued)

Table 4 (continued)

Attribute type	Constrained attributes	PDF type	# of parameter	Description
Binding site	Binding site #	Gamma	2	Min. in gamma function is determined by physical # of binding site (zero) and thus only two parameters for this gamma function: mean and stdev
Cluster size	Cluster #	Gamma	2	Min. in gamma function is determined by physical # of cluster size (one) and thus only two parameters for this gamma function: mean and stdev
Total # of parameters			28	

pdf followed by the same pdfs of SC. The fractions of SCs and ILs can be expressed as Eqs. (1) and (2):

$$\text{frac}_{SC,i} = \text{Pdf}_{SCType}(\text{Type}_i) \cdot \text{Pdf}_{SCLength}(C\#), \tag{1}$$

$$\text{frac}_{IL,i} = \text{Pdf}_{ILType}(\text{Type}_i). \tag{2}$$

Because the minimum length of SC and IL was limited by physical molecular structures, there are only two parameters for the gamma pdf to describe the length of SCs and ILs. The type of SC is limited to 4 for the example resid [32], so there are only $4-1 = 3$ parameters in the histogram pdf. Similarly, there is one parameter to adjust the two IL types.

Unlike SC and ILs, cores can be hard to describe within one continuous pdf. In addition, a single histogram pdf is also impractical to describe the core fractions in resid because there are O(100) such cores in a resid model. A practical method is to apply a juxtaposition in terms of a set of several structural attribute pdfs and obtain the fraction of cores via a joint probability calculation. Hydrocarbon building blocks could be one choice for structural attributes to constrain the fractions of cores but it is not the optimal option. The optimal structural attributes to describe the fractions of core should be strongly associated with direct or indirect measurements that were available for a heavy feedstock. For example, if applicable statistical measurements of high-resolution MS (e.g., HCHDA, FIMS, FTICR-MS) and HPLC aromatic classes were given, the fractions of compound classes, the aromatic class fractions or approximately the distribution of aromatic ring number, and the distribution of DBE or Z number could be used as the structural attributes to constrain the fractions of cores. In addition, some structural attributes derived from detailed measurements (e.g., aromatic ring number, naphthenic ring number, thiophenic ring number, etc.) were also an effective option [28–31]. Any structural attribute value of core can be treated as a property of core structure and evaluated from PropGen [30] (an in-house program in KRG to estimate thermodynamic and physical property based on the BE matrix representation). As a result, we developed a user-defined core attribute pdf sampling tool that allow users set up the attributes based on their practical situation of available analytical measurements and automatically build the equations of the joint probability calculation for cores.

A typical set of core structural attribute pdfs in a resid example [32] is shown in Table 4. We used a histogram pdf to classify the types of cores (paraffin, naphthenes, and aromatics). The naphthenic ring number and aromatic ring number were selected to control hydrocarbon fused ring structures and both followed a gamma function. A heteroatom core was limited by a conditional pdf as a binomial like the pdf shown in Eq. (3).

$$\text{frac}(x) = \text{frac}(x_{HC}) \cdot \alpha^x \cdot \binom{x_{HC}}{x} p^x (1-p)^{x_{HC}-x} \qquad (3)$$

where x is the targeted heteroatom number in a heteroatom core structure, x_{HC} is the aromatic ring number or naphthenic ring number of that core, $\text{frac}(x)$ and $\text{frac}(x_{HC})$ are the pdf values of this heteroatom group and the associated pure hydrocarbon group, p is the factor to indicate the probability of the occupancy of heteroatom in a core, and α is an adjustable factor to indicate the amount of heteroatom groups.

There were in total five heteroatom pdfs in this resid example: sulfide ring, thiophenic ring, pyridine ring, pyrrole ring, and phenol attached groups. The last pdf to constrain the fractions of cores is a ring configuration pdf that was applied as a gamma function to identify the differences among cores that were not able to be differentiated quantitatively by the above pdfs. So we applied nine structural attribute pdfs with a total of 18 parameters, as shown in Table 4, to describe the fractions of O(100) cores in this resid example.

After we acquired the fractions of cores, SCs, and ILs via selected pdfs with a limited number of parameters, we have to apply a statistical model to calculate individual molecular components' fractions based on the Bethe model scheme we discussed in the previous section. There is no evidence to suggest large-scale polymerization in petroleum, so the cluster size of a resid molecule was limited to a small number (e.g., 2 or 3). Instead of applying lattice statistics [57], we developed a simple modeling scheme. Two gamma pdfs were introduced in this scheme: the cluster size pdf and the binding site pdf shown in Table 4. Given the core and IL fractions, the fraction of a non-SC core cluster addressed before was calculated by a multinomial pdf as

$$\begin{aligned}\text{frac}_{\text{CoreCluster}} = {}&\text{Pdf}_{\text{Cluster}}(n_c) \cdot \frac{n_c!}{n_{\text{core}_i}! \cdot n_{\text{core}_j}! \cdots n_{\text{core}_k}!} \text{frac}_{\text{core}}(\text{core}_i)^{n_{\text{core}_i}} \\ &\cdot \text{frac}_{\text{core}}(\text{core}_j)^{n_{\text{core}_j}} \cdots \text{frac}_{\text{core}}(\text{core}_k)^{n_{\text{core}_k}} \\ &\cdot \frac{(n_c - 1)!}{n_{\text{IL}_m}! \cdots n_{\text{IL}_p}!} \text{frac}_{\text{IL}}(\text{IL}_m)^{n_{\text{IL}_m}} \cdot \text{frac}_{\text{IL}}(\text{IL}_p)^{n_{\text{IL}_p}}\end{aligned} \qquad (4)$$

where $\text{frac}_{\text{CoreCluster}}$ is the fraction of non-SC core cluster; $\text{Pdf}_{\text{Cluster}}(n_c)$ is the pdf value for a given cluster size n_c; $n_{\text{core}_i}, n_{\text{core}_j}, n_{\text{core}_k}$ are the number of core_i, core_j, and

$core_k$ in this cluster; $frac_{core}(core_i)$, $frac_{core}(core_j)$, $frac_{core}(core_k)$ are the fractions of $core_i$, $core_j$, and $core_k$ from core pdf; n_{IL_m} and n_{IL_p} are the number of IL_m and IL_p in this cluster; and $frac_{IL}(IL_m)$ and $frac_{IL}(IL_p)$ are the fractions of IL_m and IL_p from IL pdf.

Consequently, the mol fraction of individual component with a given binding site number via MMM method was calculated as

$$frac_{molecule} = frac_{Corecluster} \cdot Pdf_{bindingsite}(\#of\ site) \cdot Pdf_{SC}(SC_i), \quad (5)$$

where $frac_{molecule}$ is the mole fraction for a specific molecular component; $frac_{Corecluster}$ is the fraction of non-SC core cluster of this component; $Pdf_{bindingsite}$(# of site) is the pdf value of # of binding site for this component; and $Pdf_{SC}(SC_i)$ is the pdf value of SC in this component.

Finally, we applied a total of 14 pdfs with 28 parameters to build a set of joint probability mathematical equations as a quantitative sampling protocol in order to constrain the mol fractions of over 400,000 components in an example resid [32].

2.2.2 Model Optimization and Representative Results

The quantitative sampling protocol provides a unique relationship between the molecular fractions and given pdf sets. Based on it, we can employ an optimization loop to minimize an objective function that comprises a set of resid measurements, shown in Table 5. This is accomplished by adjusting the limited parameters in Table 4 until the properties based on the model composition provide a minimum value of the objective function. As a result, an optimal set of molecular compositions in a resid model was obtained.

Some representative calibration results for a resid example [32] are shown in Table 6 and Fig. 9. The prediction of general bulk properties and distillation curve shows a good agreement with experimental values. In addition, Figs. 10, 11, and 12 show the detailed molecular composition information in this resid model. From Fig. 11, DBE and carbon number distribution for pyrrolic compounds in this resid model was qualitatively consistent with the data from negative-ESI FTICR-MS. From Fig. 12, the island model structure (cluster size = 1) is the predominant in this resid model.

With the optimal molecular composition for a heavy feedstock now in hand, we can use it for the initial conditions for the reaction kinetic model analysis.

Table 5 Selected measurements for a resid example

Proper name	Measurement method	Prediction method	Weight factor	Objective function term
Density	Pycnometer	Calculated from molecular weight and molecular volume from group contribution method from Gani group	5	$\left(\dfrac{d_{cal} - d_{exp}}{5 \times d_{exp}}\right)^2$
Elemental composition	Elemental analysis	Elemental composition	0.2	$\sum_{i=1}^{\text{Element}\#} \left(\dfrac{\text{Element Wt}_{i,cal} - \text{Element Wt}_{i,exp}}{0.2 \times \text{Element Wt}_{i,exp}}\right)^2$
Molecular weight[a]	GPC	Elemental composition	5	$\left(\dfrac{Mn_{cal} - Mn_{exp}}{5 \times Mn_{exp}}\right)^2$
SARA composition	SARA analysis	Structural information[b]	5	$\sum_{i=1}^{\text{SARA}\#} \left(\dfrac{\text{SARAWt}_{i,cal} - \text{SARAWt}_{i,exp}}{5 \times \text{SARAWt}_{i,exp}}\right)^2$
Boiling point distribution	HT SimDis	Group contribution method from Gani group	1	$\sum_{i=1}^{\text{TbFrac}\#} \left(\dfrac{\text{FracWt}_{i,cal} - \text{FracWt}_{i,exp}}{1 \times \text{FracWt}_{i,exp}}\right)^2$
C-atom type	Proton NMR[c]	Atomic connectivity	0.2	$\sum_{i=1}^{\text{CType}\#} \left(\dfrac{\text{TypeMol}_{i,cal} - \text{TypeMol}_{i,exp}}{0.2 \times \text{TypeMol}_{i,exp}}\right)^2$
N-atom type	XPS	Atomic connectivity	1	$\sum_{i=1}^{\text{NType}\#} \left(\dfrac{\text{TypeMol}_{i,cal} - \text{TypeMol}_{i,exp}}{1 \times \text{TypeMol}_{i,exp}}\right)^2$
S-atom type	XPS	Atomic connectivity	1	$\sum_{i=1}^{\text{SType}\#} \left(\dfrac{\text{TypeMol}_{i,cal} - \text{TypeMol}_{i,exp}}{1 \times \text{TypeMol}_{i,exp}}\right)^2$

Reprinted with permission from [32]. Copyright (2014) American Chemical Society
[a] Only number-average molecular weights were used.
[b] Saturates were defined as molecules without aromatic ring; aromatics were defined as molecules with benzene and thiophene ring; resins and asphaltenes were summed up together as the polar class, which includes molecules with nitrogen functional groups.
[c] Data were processed by a modified Brown–Landler's method to transform 1H information to carbon type result.

Table 6 The predicted and experiment results of bulk properties in resid example [32]

Property	Predict	Exp
Density	1.1514	1.0524
Elemental (wt%)		
C	82.5	82.69
H	9.27	9.68
S	6.29	4.8
N	1	0.98
O	0.94	1.6
SARA (wt%)		
Saturates	7.1	8
Aromatics	61.6	35.4
Polars (Resin + Asp)	31.3	56.6
NMW	823	819
XPS (mol%)		
Aromatic_S_Atom	80	72
Aliphatic_S_Atom	20	28
Pyrrlic_N_Atom	59	66
Pyridinic_N_Atom	41	34
NMR		
C_A	0.38	0.36
C_N	0.19	0.19
C_P	0.43	0.45

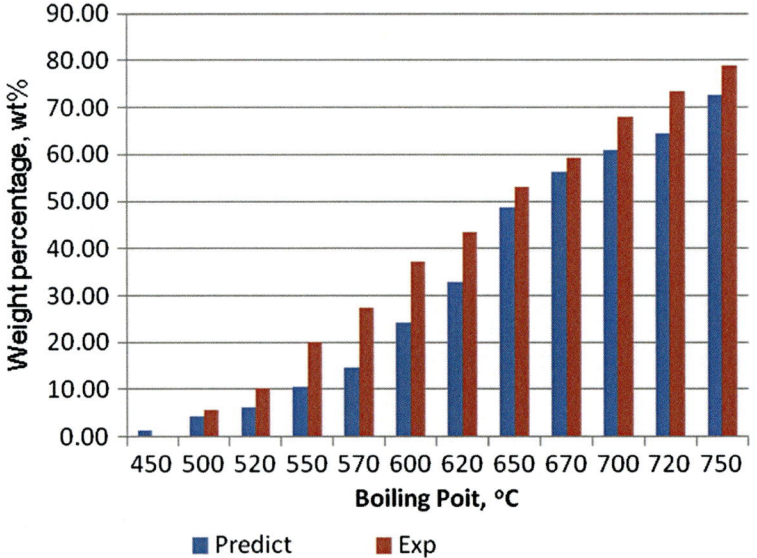

Fig. 9 The predicted and experiment results of distillation curve in resid example. Reprinted with permission from [32]. Copyright (2014) American Chemical Society

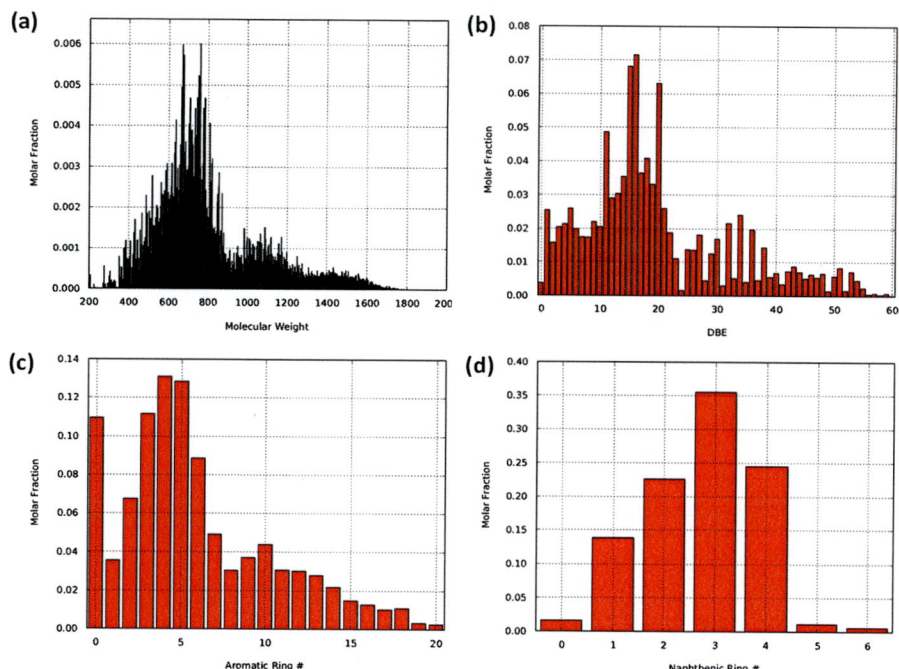

Fig. 10 Selected detailed molecular composition information in resid example. (**a**) Molecule weight distribution of a resid example; (**b**) DBE of a resid example; (**c**) aromatic ring number of a resid example; (**d**) naphthenic ring number of a resid example. Reprinted with permission from [32]. Copyright (2014) American Chemical Society

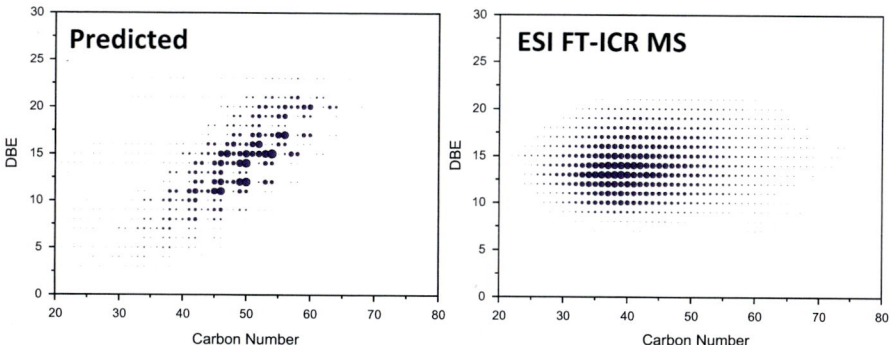

Fig. 11 DBE and carbon number distribution for pyrrolic compounds in resid example. Reprinted with permission from [32]. Copyright (2014) American Chemical Society

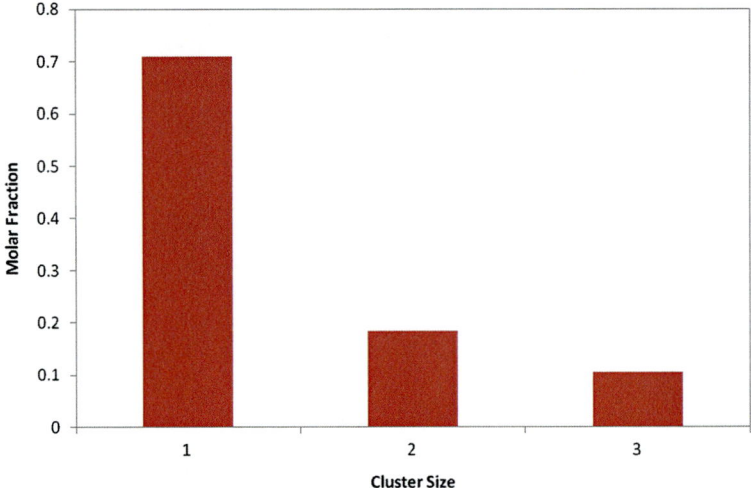

Fig. 12 Predicted cluster size distribution of in resid example. Reprinted with permission from [32]. Copyright (2014) American Chemical Society

3 Reaction Model for Heavy Oil

Coking is a resid pyrolysis refining process that converts the heavy-end fractions to a set of light-middle fractions such as gasoline, kerosene, LCO, and so on. In this article, we use resid pyrolysis as an example to show a strategy for developing a detailed kinetic model in such a complex heavy oil system.

The feedstock was selected as the example resid [32] that was discussed in the previous section. There are over 400,000 molecular components sampled in this resid sample. If we model such a feedstock via a traditional deterministic method that applies one ODE per species, it will lead to 400,000 ODEs to model the reaction system. This is far beyond the practical computational capacity we can afford today. To address this dilemma, we have developed a new approach to provide a practical number of ODEs in a reaction system that nevertheless maintains the full molecular details of the resid components. To do this, we react to the three essential structural attributes (cores, SCs, and ILs) of the resid model and then juxtapose them into molecular products at any desired point in the reactor.

3.1 ARM Reaction Network Analysis

Since all cores, SCs, and ILs were represented as atom-explicit BE matrices, we can use our in-house model building software tool called InGen to perform reaction network generation via operations on different reaction matrices. The primary reaction of resid pyrolysis is cracking. A summary of typical cracking reaction

Table 7 Cracking reaction families, sites, and matrices

Reaction type (family) and reaction site	Reaction matrix	Example reaction
Decarboxylation Carboxylic acid on side chain and irreducible molecules	H O C C H 0 -1 0 1 O -1 0 1 0 C 0 1 0 -1 C 1 0 -1 0	
Naphthenic ring opening Six-membered naphthenic rings	C C C H C 0 -1 0 1 C -1 0 1 0 C 0 1 0 -1 H 1 0 -1 0	
Sulfide ring opening Five-membered sulfide rings	H C C S H 0 -1 0 1 C -1 0 1 0 C 0 1 0 -1 S 1 0 -1 0	
Thermal cracking: hydrocarbon Hydrocarbon side chains and irreducible molecules	C C C H C 0 -1 0 1 C -1 0 1 0 C 0 1 0 -1 H 1 0 -1 0	
Thermal cracking: C–S bonds Carbon–sulfur bonds on side chains and irreducible molecules	H C C S H 0 -1 0 1 C -1 0 1 0 C 0 1 0 -1 S 1 0 -1 0	

Reprinted with permission from [33]. Copyright (2015) American Chemical Society

families was shown in Table 7. Aromatization and ring condensation, shown in Table 8, are two other secondary reactions considered in a resid pyrolysis.

The cores, SCs, and ILs of the feedstock resid model [32] were input as the seed to InGen, which performed the corresponding reactions on those attributes. A summary of these results is shown in Table 9 [33]. In this example, the number of cores was extended from 373 to 1,704 and the number of SCs was extended from 200 to 256. In addition, 876 irreducible small molecules (IM) were introduced to the system. So, in total, we obtained 2,839 species (and therefore ODEs) participating in 6,274 reactions for this resid model. This provided the reaction kinetics for the 400,000 molecules, representing a desirable and practical reduction in the computational demand.

Table 8 Aromatization and ring condensation reaction families, sites, and matrices

Reaction type (family) and reaction site	Reaction matrix	Example reaction
Naphthenic ring aromatization-2	<table><tr><td></td><td>C</td><td>C</td><td>H</td><td>H</td></tr><tr><td>C</td><td>0</td><td>1</td><td>-1</td><td>0</td></tr><tr><td>C</td><td>1</td><td>0</td><td>0</td><td>-1</td></tr><tr><td>H</td><td>-1</td><td>0</td><td>0</td><td>1</td></tr><tr><td>H</td><td>0</td><td>-1</td><td>1</td><td>0</td></tr></table>	
Naphthenic ring aromatization-4	<table><tr><td></td><td>C</td><td>C</td><td>C</td><td>C</td><td>H</td><td>H</td><td>H</td><td>H</td></tr><tr><td>C</td><td>0</td><td>1</td><td>0</td><td>0</td><td>-1</td><td>0</td><td>0</td><td>0</td></tr><tr><td>C</td><td>1</td><td>0</td><td>0</td><td>0</td><td>0</td><td>-1</td><td>0</td><td>0</td></tr><tr><td>C</td><td>0</td><td>0</td><td>0</td><td>1</td><td>0</td><td>0</td><td>-1</td><td>0</td></tr><tr><td>C</td><td>0</td><td>0</td><td>1</td><td>0</td><td>0</td><td>0</td><td>0</td><td>-1</td></tr><tr><td>H</td><td>-1</td><td>0</td><td>0</td><td>0</td><td>0</td><td>1</td><td>0</td><td>0</td></tr><tr><td>H</td><td>0</td><td>-1</td><td>0</td><td>0</td><td>1</td><td>0</td><td>0</td><td>0</td></tr><tr><td>H</td><td>0</td><td>0</td><td>-1</td><td>0</td><td>0</td><td>0</td><td>0</td><td>1</td></tr><tr><td>H</td><td>0</td><td>0</td><td>0</td><td>-1</td><td>0</td><td>0</td><td>1</td><td>0</td></tr></table>	
Naphthenic ring aromatization-6	6-Carbon, 6-hydrogen analogue to reaction matrix in naphthenic ring aromatization-4	
Sulfide ring aromatization-3	Same matrix as naphthenic ring aromatization-2	
Sulfide ring aromatization-5	Same matrix as naphthenic ring aromatization-4	
Aromatic ring condensation		

Reprinted with permission from [33]. Copyright (2015) American Chemical Society

3.2 ARM Model Equation

From the InGen results, the 2,828 ODE equations of this model can be written in terms of three attributes (core, SC, and IL) plus irreducible molecules (IM), respectively:

Table 9 The statistics of reaction network info for a resid pyrolysis model [32]

Reaction family	# of reactions
Decarboxylation	137
Naphthenic ring opening	224
Sulfide ring opening	9
Thermal cracking: hydrocarbon	3,790
Thermal cracking: C–S bonds	1,097
Naphthenic ring aromatization-6	8
Naphthenic ring aromatization-4	734
Naphthenic ring aromatization-2	13
Sulfide ring aromatization-5	9
Sulfide ring aromatization-3	6
Aromatic ring condensation	1
# of total reactions	6,274
Species information	
# of cores	1,704
# of IL	3
# of SC	256
# of IM	876
# of total species	2,839

$$\frac{dCore}{dt} = \sum_{i,\text{reactions}} v_i \cdot \text{rate}_i$$
$$\frac{dSC}{dt} = \sum_{i,\text{reactions}} v_i \cdot \text{rate}_i$$
$$\frac{dIL}{dt} = \sum_{i,\text{reactions}} v_i \cdot \text{rate}_i \qquad (6)$$
$$\frac{dIM}{dt} = \sum_{i,\text{reactions}} v_i \cdot \text{rate}_i$$

Those four categories of ODEs were integrated to an executable code and solved simultaneously via the in-house software KME [30] after the kinetic rate parameters were specified.

3.3 Kinetic Parameters: LFER

The reaction family and LFER concepts were applied to determine the kinetics of this resid pyrolysis model. For any reaction family, the rates of reactions within this family can be written as Eq. (7)

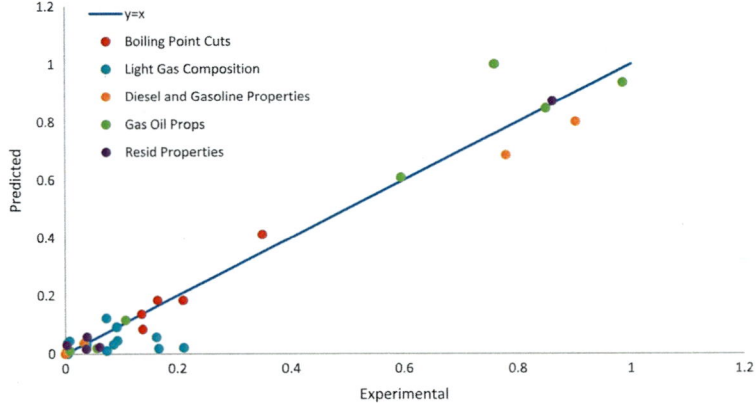

Fig. 13 The parity plot between predictive values and experimental values for a resid pyrolysis model. Reprinted with permission from [33]. Copyright (2015) American Chemical Society

$$\mathrm{Ln}\, k_i = \mathrm{Ln} A - \frac{E}{RT}, \quad (7)$$

$$E = E_\circ + \alpha \cdot \Delta H_{\mathrm{rxni}}.$$

So there are only three parameters per reaction family in a reaction model. The resid pyrolysis model [33] has 11 reaction families, and the LFER approach was applied to 10 of them. The aromatic condensation reaction was constrained by a separate correlation expression [33].

3.4 Post-reaction Sampling and Product Property Estimation

The post-reaction results were expressed as product attributes that were then converted to molecular components in order to calculate the product properties. The algorithm for sampling product attributes back to molecular components is the same as the quantitative sampling protocol of feedstocks we discussed in the previous section. Notice that all post-reaction fractions of cores, SC, and ILs were derived directly from the kinetic calculation, so it is a straightforward sampling process to convert the attribute values to product molecular compositions.

The product properties and yields were calculated based on those sampled molecular compositions from post-reaction attributes (cores, SCs, and ILs) and IMs. As a result, an objective function can be written in terms of selected product properties in order to tune the kinetic parameters via an optimization loop.

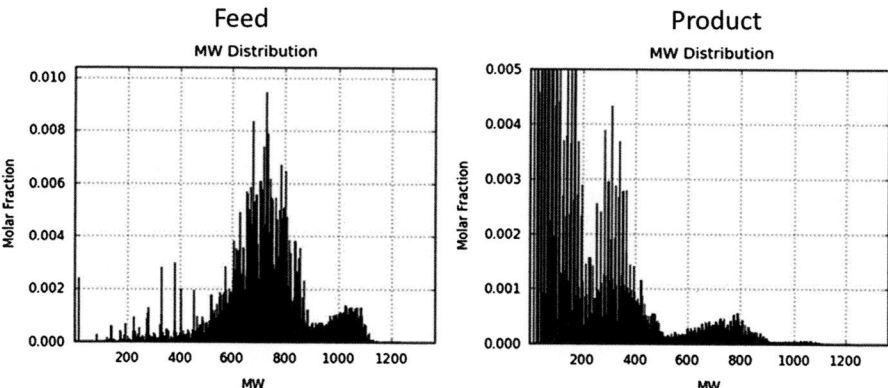

Fig. 14 Selected molecular details of resid pyrolysis model. Reprinted with permission from [33]. Copyright (2015) American Chemical Society

3.5 Representative Results of a Resid Pyrolysis Model

The representative tuning results [33] are shown in Fig. 13. The distillation curve and liquid product yields were used as calibration items in this example. Figure 13 shows the predicted values are in good agreement with the experimental values. Selected molecular details of this example are shown in Fig. 14. Figure 14 shows the comparison of molecular weight distribution before and after resid pyrolysis. After pyrolysis, the major peak of this distribution was shifted to low molecular weight (200–400) with a peak remaining in the higher molecular weight range due to the coking reaction.

4 Summary

In order to represent the molecular compositions of a heavy feedstock, we describe it in terms of three essential structural attributes (cores, SCs, and ILs). From the state-of-the-art analytical measurements, the molecular attributes of cores, SCs, and ILs can be obtained and juxtaposed into 400,000 representative components. Those attributes can be constrained by a set of pdfs. By applying an optimization loop, we can adjust the 28 associated parameters to match the objective function in terms of available measurements and thus get an optimal set of molecular compositions for a heavy feedstock.

The molecular compositions of a feedstock serve as the initial conditions for a reaction model. Instead of modeling over 400,000 molecules directly, we applied the ARM model approach to perform the reaction model in terms of 2,839 attributes and equations. This allowed us to model the full molecular detail with almost two orders of magnitude reduction in the number of model equations. The representative tuning results confirmed the validity of this approach.

References

1. Froment GF (2005) Single event kinetic modeling of complex catalytic processes. Catal Rev Sci Eng 47(1):83–124
2. Froment GF et al (2006) Alkylation on solid acids. Part 2. Single-event kinetic modeling. Ind Eng Chem Res 45:954–967
3. Froment GF et al (1993) Single-event kinetics of catalytic cracking. Ind Eng Chem Res 32:2997–3005
4. Evans MG, Polanyi M (1938) Inertia and driving force of chemical reactions. Trans Faraday Soc 31:11
5. Quann RJ, Jaffe SB (1992) Structure oriented lumping. Describing the chemistry of complex hydrocarbon mixtures. Ind Eng Chem Res 31(11):2483–2497
6. Quann RJ, Jaffe SB (1996) Building useful models of complex reaction systems in petroleum refining. Chem Eng Sci 51(10):1615
7. Jaffe SB, Freund H, Olmstead WN (2005) Ind Eng Chem Res 44:9840
8. Peng B, Towler G (Supervisor) (1999) Molecular modelling of petroleum processes. Ph.D Dissertation, University of Manchester Institute of Science and Technology, Manchester, pp 22–41
9. Hu S, Zhu XX (2001) Appl Therm Eng 21:1331
10. Mi Saine Aye M, Zhang N (2005) Chem Eng Sci 60:6702
11. Gomez-Prado J, Zhang N, Theodoropoulos C (2008) Energy 33:974
12. Wu Y, Zhang N (2009) Chem Eng Trans 18:749–754
13. Wei W (2004) The interface of chemical engineering and it in kinetics models. Doctoral Dissertation, Rutgers University
14. Hudebine D, Verstraete JJ (2004) Chem Eng Sci 59:4755
15. Hudebine D, Verstraete J, Chapus T (2009) Oil Gas Sci Technol 66:461–477
16. Verstraete JJ, Schnongs P, Dulot H, Hudebine D (2010) Chem Eng Sci 65:304
17. de Oliveira LP, Vazquez AT, Verstraete JJ, Kolb M (2013) Energy Fuel 27:3622
18. Broadbelt LJ, Stark SM, Klein MT (1996) Comput Chem Eng 20(2):113–129
19. Joshi PV (1998) Molecular and mechanistic modeling of complex process chemistries. Ph.D Dissertation, University of Delaware
20. Gang H (2001) Integrated chemical engineering tools for the building, solution, and delivery of detailed kinetic models and their industrial applications. Doctoral Dissertation, University of Delaware
21. Bennett C (2010) User-controlled kinetic network generation with INGen. Doctoral Dissertation, Rutgers University
22. Watson BA, Klein MT, Harding RH (1996) Mechanistic modeling of n-heptane cracking on HZSM-5. Ind Eng Chem Res 35:1506–1516
23. Watson BA, Klein MT, Harding RH (1997) Catalytic cracking of alkylbenzenes: modeling the reaction pathways and mechanisms. Appl Catal A Gen 160:13–39
24. Watson BA, Klein MT, Harding RH (1997) Mechanistic modeling of n-hexadecane cracking on rare earth Y. Energy Fuels 11:354–363
25. Watson BA, Klein MT, Harding RH (1997) Mechanistic modeling of a 1-phenyloctane/n-hexadecane mixture on rare earth Y zeolite. Ind Eng Chem Res 36:2954–2963
26. Watson BA, Klein MT, Harding RH (1997) Catalytic cracking of alkylcyclohexanes: modeling the reaction pathways and mechanisms. Int J Chem Kinet 29(7):545–560
27. Klein MT et al (2005) Molecular modeling in heavy hydrocarbon conversions. CRC, Boca Raton. ISBN 978-0-8247-5851-6
28. Trauth DM (1993) Structure of complex mixtures through characterization, reaction, and modeling. Ph.D Dissertation, University of Delaware
29. Campbell DM (1998) Stochastic modeling of structure and reaction in hydrocarbon conversion. Doctoral Dissertation, University of Delaware
30. Hou Z (2011) Ph.D Dissertation, Rutgers University, New Jersey

31. Pyl SP, Hou Z, Van Geem KM, Reyniers MF, Marin GB, Klein MT (2011) Ind Eng Chem Res 50:10850
32. Zhang L, Hou Z, Horton SR, Klein MT, Shi Q, Zhao S, Xu C (2014) Molecular representation of petroleum vacuum resid. Energy Fuel 28:1736–1749
33. Horton SR, Hou Z, Moreno BM, Bennett CA, Klein MT (2013) Molecule-based modeling of heavy oil. Sci China Chem 56:840–847
34. Horton SR, Zhang L, Hou Z, Bennett CA, Klein MT et al (2015) Molecular-level kinetic modeling of resid pyrolysis. Ind Eng Chem Res 54(16):4226–4235
35. Boduszynski MM (1988) Energy Fuel 2:597
36. Kendrick E (1963) Anal Chem 35(13):2146–2154
37. Hughey CA et al (2002) Org Geochem 33:743–759
38. Qian K, Edwards KE et al (2007) Energy Fuel 21:1042
39. Smith DF, Rahimi P, Teclemariam A, Rodgers RP, Marshall AG (2008) Energy Fuel 22:3118
40. Zhang L, Xu Z, Shi Q, Sun X, Zhang N, Zhang Y, Chung KH, Xu C, Zhao S (2012) Energy Fuel 26:5795
41. Zhang T, Zhang L, Zhou Y, Wei Q, Chung KH, Zhao S, Xu C, Shi Q (2013) Energy Fuel 27:2952
42. Shi Q, Hou D, Chung KH, Xu C, Zhao S, Zhang Y (2010) Energy Fuel 24:2545
43. Shi Q, Pan N, Liu P, Chung KH, Zhao S, Zhang Y, Xu C (2010) Energy Fuel 24:3014
44. Podgorski DC, Corilo YE, Nyadong L, Lobodin VV, Robbins WK, McKenna AM, Marshall AG, Rodgers RP (2013) Energy Fuel 27(3):1268–1276
45. McKenna AM, Donald LJ, Fitzsimmons JE, Juyal P, Spicer V, Standing KG, Marshall AG, Rodgers RP (2013) Energy Fuel 27(3):1246–1256
46. McKenna AM, Blakney GT, Xian F, Glaser PB, Rodgers RP, Marshall AG (2010) Energy Fuel 24:2939
47. McKenna AM, Purcell JM, Rodgers RP, Marshall AG (2010) Energy Fuel 24:2929
48. McKenna AM, Marshall AG, Rodgers RP (2013) Energy Fuel 27(3):1257–1267
49. Qian K, Edwards KE, Mennito AS, Freund H, Saeger RB, Hickey KJ, Francisco MA, Yung C, Chawla B, Wu C, Kushnerick JD, Olmstead WN (2012) Determination of structural building blocks in heavy petroleum systems by collision-induced dissociation Fourier transform ion cyclotron resonance mass spectrometry. Anal Chem 84:4544–4551
50. Marshall AG, Rodgers RP (2004) Petroleomics: the next grand challenge for chemical analysis. Acc Chem Res 37:53–59
51. Benson SW (1968) Thermochemical kinetics. Wiley, New York, Chap. 2
52. Hukkerikar AS et al (2012) Fluid Phase Equilib 321:25–43
53. Zhang ZG, Guo S, Zhao S, Yan G, Song L, Chen L (2008) Energy Fuel 23:374
54. Su Y, Artok L, Murata S, Nomura M (1998) Energy Fuel 12:1265
55. Vanderzande C (1998) Lattice models of polymers, Cambridge lecture notes in physics 11. Cambridge University Press, Cambridge
56. Grant DM, Pugmire RJ (1989) Energy Fuel 3(2):175–186
57. Peng PA, Fu J, Sheng G, Morales-Izquierdo A, Lown EM, Strausz OP (1999) Energy Fuel 13:266

Molecular Modeling for Petroleum-Related Applications

Liang Zhao, Dong Zhai, Huimin Zheng, Jingjing Ji, Lei Wang, Shiyi Li, Qing Yang, and Chunming Xu

Abstract Over the last few decades, researchers have extensively studied the conversion of fossil resources into fuels and chemicals. Nearly all transportation fuels are produced by a series of catalytic processes. For example, about 70–80 wt% of the motor gasoline in China is produced through fluid catalytic cracking (FCC) by zeolitic catalysts. Petroleum chemical reactions contain complicated reaction networks and complicated reaction pathways. Under certain conditions, the types and directions of chemical reactions could be determined by the characteristics of the feedstock and the properties of the catalysts used. Therefore, novel catalysts with high activities and selectivities are required for special production aims. Molecular modeling is a useful tool for catalyst design and provides insights into the catalytic reaction processes at the atomic or molecular level. Modeling helps us fully understand the microscopic mechanisms and identify the key factors that affect the structure–activity relationships. Moreover, with the rapid development of computer techniques and theoretical methods, bigger systems and more complicated reaction processes, such as the adsorption and diffusion behaviors of adsorbates in the zeolite pore system by molecular simulation or the catalytic reaction mechanism by quantum chemical calculations, can be revealed at the molecular level. This work reviews some applications of molecular modeling methods and attempts to design novel catalysts for petroleum refining.

Keywords Adsorption • Diffusion • Molecular modeling • Pore structure • Reaction • Zeolite

---

L. Zhao, D. Zhai, H. Zheng, J. Ji, L. Wang, S. Li, Q. Yang, and C. Xu (✉)
State Key Laboratory of Heavy Oil Processing, China University of Petroleum (Beijing), 18 Fuxue Road, Beijing 102249, China
e-mail: Liangzhao@cup.edu.cn; xcm@cup.edu.cn

Contents

1	Introduction	122
2	Theory and Methods	123
	2.1 Classification	123
	2.2 QM-Based Methods	123
	2.3 Force-Field-Based Methods	125
	2.4 MM Models	127
	2.5 MC Methods	127
	2.6 MD Methods	128
3	Applications in Heavy Oil Chemistry	128
	3.1 Improved B–L Method	130
	3.2 Most Probable Molecular Conformation of Heavy Oil	132
4	Applications in Catalyst Development	135
	4.1 Adsorption	135
	4.2 Diffusion	143
	4.3 Catalyst Structure	149
5	Applications in Reactions of Refining Processes	152
	5.1 Desulfurization	152
	5.2 Hydrodenitrogenation (HDN)	160
	5.3 Alkylation	163
	5.4 Isomerization	166
	5.5 Hydrodeoxygenation (HDO)	169
6	Summary and Prospects	170
References		170

1 Introduction

Rapid improvements in theoretical methods, computer speed, computer memory, and algorithm efficiency have improved efforts to model physical and chemical processes, thereby widening the applications of molecular modeling in numerous fields. For example, modeling has been used to predict the behavior of properties under unusual conditions, such as high pressure and temperature, that describe the processes that cannot be observed by current experimental approaches, simulate reaction processes that cannot be performed in the laboratory, and test the limiting conditions of various empirical laws. Molecular modeling offers a way to understand the invisible microscopic world and draws a meaningful and colorful picture to answer the questions raised by experiments. In addition, molecular modeling saves both time and money when used as a tool to predict or suggest the direction of an experiment during the design of experimental schemes.

Specifically in the development of high-performance computing (HPC) techniques, molecular modeling has been widely applied in considerably large petroleum systems to determine the structure–activity relationships between large molecules and catalysts, the reaction networks and mechanisms in petroleum processing and production refining, and the pore structure design of zeolites.

This paper presents an introduction and overview of the chemistry of petroleum refining, including modeling theories and methods, main applications in petroleum refining, and our group's contributions to these studies.

2 Theory and Methods

2.1 Classification

Theory and methods may be grouped according to different standards of classification. For example, these methods can conveniently classify models into four groups based on their length and time scales [1]: (a) the *electronic scale*, where matter is considered to be made up of fundamental particles and described by quantum mechanics; (b) the *atomistic level*, where matter is made up of atoms with behaviors obeying the laws of statistical mechanics; (c) the *mesoscale level*, where matter is composed of blobs of matter, each of which contains a number of atoms; and (d) the *continuum level*, where matter is regarded as a continuum and the well-known macroscopic laws apply. However, a simpler classification scheme is adopted in the present paper. The modeling methods described in this study are classified into two groups depending on whether a force field is involved or not; thus, methods are classified as quantum mechanics (QM)-based methods or force-field-based methods. The QM-based methods include the ab initio methods, the semiempirical approach, and the density functional theory (DFT) [2]. The force-field-based methods mainly include molecular mechanics (MM), the Monte Carlo method (MC), and molecular dynamics (MD). Other methods, such as Brownian dynamics and dissipative particle dynamics, are not discussed in this paper.

2.2 QM-Based Methods

QM assumes that all of the actions of electrons can be represented by a wave function described by the Schrödinger equation:

$$H\Psi = E\Psi, \tag{1}$$

where H is the Hamiltonian operator, Ψ is the wave function, and E is the energy. QM can calculate the probability of the appearance of electrons. The probability density distribution is $|\psi|^2$.

QM is a precise mathematical description of the electronic behaviors within atoms. In theory, this method can predict any property of an individual atom or molecule. However, the QM equations have only been solved exactly for one-electron systems. Therefore, several other QM-based methods have been developed to approximate the solutions for multiple electron systems and predict molecular properties ranging from geometries to energetics to the distribution of electron densities. In particular, QM may be used to investigate the chemical reactions of bond breaking and formation.

2.2.1 Ab Initio Methods

By taking the Planck constant, electron mass, quantity of electricity, and atomic number of an element into consideration, the Shrödinger equation is solved without any empirical constants, and the molecular integral of the whole system is calculated. The most common type of ab initio calculation is called the Hartree–Fock calculation (HF).

Ab initio calculations are associated with the basis set of atomic orbitals, which are used to describe the molecular orbitals. These basis sets are usually correlated with shells, such as s or sp shells. A minimal basis set has a rather limited variational flexibility, particularly if the exponents are not optimized. The first step to improve the minimal basis set is the so-called split-valence basis set, wherein two basic functions are used for each valence atomic orbital. Obviously, more computational time is required by using the split-valence technique than by using the minimal basis set for the same molecular system. However, a considerable increase in the computational efficiency can be achieved if the exponents of the Gaussian primitives are shared between different basis functions. At the split-valence level, primitive exponents are shared between s and p functions for the valence shells. In particular, a series of basis sets at the split-valence level have been defined and designated N-31G, where N stands for the integer number of the Gauss function. The next step to improve a basis set could be a triple zeta or quadruple zeta. This step usually involves addition of polarization functions. Using standard notation, a basis set with a single asterisk (*) indicates addition of a d-type function to heavy atoms (atomic numbers greater than 2) and double asterisks (**) indicate addition of d-type functions to heavy atoms and p-type functions to light atoms (H and He).

2.2.2 Semiempirical Methods

Ab initio calculations are useful and accurate. However, the method is tedious and time consuming, especially for large molecular systems. Thus, semiempirical quantum mechanical methods have become very powerful and attractive to many chemists and biochemists because of their simplicity and quick computational speed. Semiempirical methods are based on the wave function, Hamilton factor, and integral which can be applied for predicting various chemical properties, such as molecular structures and conformations, heats of formation, ionization potentials, and electron affinities. The methods are fit for organic compounds and can obtain significantly accurate calculation results of macromolecules. Conventional semiempirical methods involve traditional parameters, which are parameterized to yield good molecular structures and/or heats of formation values. The accuracy of semiempirical QM methods depends on the database used to parameterize the method. An advantage for specific and well-parameterized molecular systems is that they calculate values closer to the experimental data than the lower level ab initio and DFT techniques. A disadvantage of these methods, however, is that relevant parameters must be available before running a calculation. Developing appropriately accurate parameters is time consuming.

2.2.3 DFT

DFT is one of the most widely used computational approaches. It considers a relationship between the electronic energy and density similar to the Thomas–Fermi model. DFT indicates that the ground-state energy of a system can be decided by its electronic density. The energy function is written as follows:

$$E(\rho) = T(\rho) + U(\rho) + E_{xc}(\rho), \qquad (2)$$

where E is the total energy of a system, T is the kinetic energy of electrons, U is the classical potential energy of coulomb, and E_{xc} is the exchange correlation potential. The precision of DFT depends on the quality of E_{xc}. Theory developers often endeavor to find, design, and propose an appropriate E_{xc} to enhance the precision and reliability of DFT.

The basis sets in QM are used to describe electrons and express wave functions, which consist of atomic functions. The basic methods contain (a) atomic orbital basis sets (e.g., Slater type, Gaussian type, SVP, TZP, QZP, cc-pVXZ, etc.), (b) a plane-wave pseudopotential method, and (c) wavelets, among others. Plane-wave methods have some advantages: (a) A real-space representation means the potential energy V is relatively diagonal, whereas a momentum-space representation means the kinetic energy T is diagonal. Plane-wave methods can easily change from one to another. (b) It can be neglected to control basis-set convergence. (c) Superposition errors in plane-wave methods can be avoided and must be carefully considered in calculations based on local basis sets. However, the treatment of exact exchange in these methods is difficult. Pseudopotentials are necessary for all plane-wave methods and can also be used in local basis-set methods. Pseudopotentials can be used to avoid treating strongly bound and chemically inert core electrons.

2.3 *Force-Field-Based Methods*

In molecular mechanics simulations, force fields are mathematical functions used to describe the potential energy of a statistical mechanical model made up of a system of molecules or atoms. The accuracy of the simulation depends on the quality of the force field. The basic functional form of a force field comprises bonded terms related to atoms that are linked by covalent bonds and non-bonded terms describing long-range electrostatic and van der Waals' forces:

$$E = E_{\text{bonded}} + E_{\text{nonbonded}}, \tag{3}$$

where E is the total potential energy of a system, E_{bonded} is the bonded interaction energy, and $E_{\text{nonbonded}}$ is the non-bonded interaction energy. E_{bonded} can be described as

$$E_{\text{bonded}} = E_{\text{bond}} + E_{\text{angle}} + E_{\text{torsion}}, \tag{4}$$

where E_{bond} is the bond-stretching energy, E_{angle} is the angle-bending energy, and E_{torsion} is the torsional energy. $E_{\text{nonbonded}}$ can be represented by

$$E_{\text{nonbonded}} = E_{\text{vdw}} + E_{\text{Coulomb}}, \tag{5}$$

where E_{vdw} is the van der Waals' item and E_{coulomb} is the Coulomb item.

In the potential functional form, the force field also defines a set of parameters for each type of atom. A force field would include distinct parameters for each atom type in different chemical environments. A typical parameter set contains the atomic mass, van der Waals' radius, the partial charge of each atom type, the force constants for each potential, and the equilibrium values of the bond lengths, bond angles, and dihedral angles for pairs, triplets, and quadruplets of bonded atoms. Parameterization is very important because the performance of a force field is sensitive to the parameters. To create a force field, the data for parameterization must include the geometry and conformational energy of key molecules and their vibrational frequencies. In some cases, experimental data is difficult to obtain or is even nonexistent. Consequently, QM calculations are used to present the data so that more systems can be treated with the force-field method. After the data are determined, parameterization is achieved through two approaches: by trial and error and by the least-squares method.

An important feature of force fields is the transferability of parameters and functional forms. Therefore, the same set of parameters can be used for a series of related molecules. Otherwise, individual force fields would be impossible to create because too many parameters will be required. When parameters are transferred from one force field to another, some of them may become invalid. Scholars must try to determine this information by using programs that can automatically predict parameters or atomic properties.

Force fields have been developed for several years. As systems have become more complex, over a hundred force fields have been created for different purposes, including AMBER, CHARM, CVFF, CFF, UFF, and so on. Our work introduced two commonly used force fields to the petroleum field. The first of these force fields is the MM-type force field, which was originally developed by Peter Kollman's group at the University of California, San Francisco. This approach classifies carbon atoms into different types with different parameters. The MM-type force field is very suitable for organic compounds. The configurations, thermodynamic properties, vibrational spectra, and lattice energies obtained from this force field are precise because it considers crossing interaction terms. The second force field is the COMPASS force field, which has a form similar to that of CFF91 but requires rigorous parameterization and even includes some ab initio calculations. The COMPASS force field is important because it presents several parameters of metal and metallic oxide. These parameters give rise to a high-quality force field that can also be used for inorganic systems. This model can integrate organic and inorganic parameters, which makes it suitable for metallic oxide and metal-organic compounds.

2.4 MM Models

The MM approach is dependent on the available force fields. According to the Born–Oppenheimer approximation, MM ignores the movement of electrons and considers the energy of a system as a function of the nuclear positions. This simple model is used to simulate processes like the stretching of bonds or the opening and closing of angles. MM features a number of specific methods, such as lattice statics and lattice dynamics for solid-state systems. The most commonly used method is energy minimization. A number of minimum points are present on the energy surface, which correspond to the stable state of a system. Any other position has a higher energy. Among these states, the position with the lowest energy is called the global energy minimum. A minimization algorithm is used to determine how the system changes from one minimum energy to another. Standard methods are used to find the minimum of an analytical function. However, for molecular systems where the energy varies with the coordinates, we can only use numerical methods, which determine the minimum by gradually changing the coordinates to produce configurations with lower and lower energies. MM ignores the motion of electrons; thus, this method cannot determine properties related to electronic distribution.

2.5 MC Methods

MC methods are a class of computational algorithms that depend on random sampling to obtain numerical results. These methods first randomly generate a configuration and then use special criteria to decide whether to accept the new result. The Boltzmann factor is used to evaluate the probability of accepting the configuration, which is calculated by the potential energy function. If the resulting energy is lower than that of its predecessor, the new configuration is accepted. If the resulting energy is higher, the difference between Boltzmann factors is calculated and a random number between 0 and 1 is selected. When compared with the difference of the random number, two results are obtained. If the number is higher than the difference, the new configuration is rejected, and the old one will be used for the next iteration. If the number is lower, the new configuration will become the next state.

In MC computations, a configuration does not depend on any other configuration but its predecessor. A new configuration may be generated or obtained by moving atoms or molecules or by rotating bonds. The total energy is decided by the potential energy function alone. The model generally obtains samples from the canonical ensemble (constant N or V and temperature, T). MC is usually used to investigate the structures and properties of phase changes of complex systems and can calculate statistical average values but not acquire timely dynamic information.

2.6 MD Methods

MD is the computer simulation of physical movements of a system that consists of atoms and molecules. In general, the trajectories of atoms and molecules are determined by numerically integrating the Newton's equations of motion for a system of interacting particles, where forces between the particles and potential energy are defined by force fields. Initially, a system of N particles is abstracted into N mass points that interact with each other; the trajectories of motion of these particles are considered functions of their positions and momentum. MD calculates the velocity and acceleration of these particles from Newton's equation of motion, such that the new positions and velocity of the points are obtained at a time step. Thus, a trajectory of motion of the atoms may be predicted with a given number of time steps. Finally, we analyze the trajectory by using statistical methods to obtain the dynamic and thermodynamic properties of the system. In several aspects, MD is mainly applied to simulations of equilibrium states; the method can calculate the dynamic properties of systems and predict future states from the current state.

3 Applications in Heavy Oil Chemistry

Petroleum is a complex mixture composed of several kinds of chemicals. Heavy oil is the most complex constituent of petroleum because this component presents a higher molecular weight (MW) and a more complicated composition and molecular configuration. The term "heavy" indicates that the boiling points of the said components are higher than 650°F (345°C), for example, distillation residues. In order to study the great variety of petroleum components, we extensively use the concept of atmospheric equivalent boiling point (AEBP). The AEBP scale encompasses the entire boiling range of petroleum, including the equivalent boiling range of non-distillable residue fractions. Because the composition information of raw material is necessary for selecting appropriate processing route or for improving and optimizing existing processes, knowledge of the detailed distributions of high-boiling, residual fractions (i.e., the AEBP distribution curve) and concentration of heteroatoms are required; moreover, the concentration and types of heterocompounds and the carbon number distribution of these compound types must also be identified. However, availability of compositional information rapidly decreases with increasing boiling point. To recognize and utilize these resources sufficiently, the chemical compositions and structures of molecules should be better understood.

Previous studies have shown that heavy oil contains five basic elements, namely, C, H, O, N, and S. Trace metal elements, such as Ni, V, and Cu, also exist in heavy oil. Experimentally, we can analyze the elemental composition of heavy oil. In particular, the hydrocarbon ratio of the atomic number (H/C) can provide crucial information with which to characterize average structures or group contents. If the molecules contain ring structures, the H/C decrease; when the molecules contain polycyclic aromatic hydrocarbons, the H/C decreases apparently. Besides the hydrocarbon ratio, another parameter Z from mass spectrometry is related to the chemical structure. Z is defined by molecular formula ($C_nH_{2n+Z}X$), which indicates the index of hydrogen deficiency comparing with paraffin molecules (for paraffin $Z=2$); X is expressed as a heteroatom. The value of Z can be calculated by the formula: $Z = -2[(R + n_{DB}) - 1]$, where R is the total ring number and n_{DB} is the number of double bonds.

Hetero-compounds, particularly sulfur compounds, can be found in almost all petroleum fractions, even those with lower boiling points. The sulfur concentration increases moderately and in roughly linear fashion with the boiling point. Consequently, much of the sulfur in crude oil resides in the distillates. Nitrogen and oxygen have much lower concentrations in low-boiling point fractions than in high-boiling point ones; these concentrations also initially increase moderately with increasing boiling point until near 650°F (345°C). Metals are nearly absent in distillates below 1,000°F (540°C). On the other hand, in higher-boiling fractions, these heterocompounds are difficult to detect on account of their greater complexity of ring structure as well as number and length of substituents. As well, fractions with higher AEBPs comprise increasing amounts of compounds with more heteroatoms. Therefore, the AEBP theory provides a rational basis for comparing different crude oils as well as feeds, intermediates, and products from various refining processes.

Statistically speaking, this thought is creative because it regards complex mixtures as a composition of average molecules with the same series of structures, and it considers that the average molecular is made up of several structural units. However, some scholars are opposed to characterizing heavy oil in terms of average structures because they believe these structures do not represent the actual structures. No ideal characterization method for such complex systems is yet available. Thus, research on heavy oil chemical structures mostly adopts the average structural parameters method.

During average structure determination of crude oil vacuum distillates, the n-d-M method is commonly used. This method is an extension of early direct methods and the Waterman [3] method, which was systematically summarized by Van Nes and Van Westen [4]. In practical applications, several refining experiments used the method to obtained hundreds of precise data, after analysis, the method was found to present high accuracy but two limitations: 1) The average molecular structure of the sample oil cannot have too many rings. The ring system must be cata-condensed with no peri-condensed regions, and the parameters $RT \ngtr 4$, $RA \ngtr 2$, $CR \ngtr 75\%$, and $CA/CN \ngtr 1.5$ must be preserved; 2) the heteroatom content (mass fraction) in the sample oil cannot be too high, and the following parameters $S \ngtr 2\%$, $N \ngtr 0.5\%$, and $O \ngtr 0.5\%$ must be preserved. Modern physical instrumental analytical methods, such as nuclear magnetic resonance (NMR), infrared spectroscopy (IR), mass spectroscopy (MS), and paramagnetic resonance spectroscopy (ESR), among others, can be combined. Brown and Ladner developed the Brown–Ladner method (B–L method), which was the first to propose three average structural parameters [5]. This parameter is still commonly used to study the chemical structures of heavy oil. Combined with more experimental data on the latter, this method has been further extended to other applications, and considerable research has been achieved by using quantum chemistry calculations and molecular simulation methods.

3.1 Improved B–L Method

The B–L method is mainly based on elemental compositions, with various types of hydrogen fractions determined through NMR (H-NMR) as the raw data. The improved B–L method adds the relative molecular weight, IR, and C-NMR as raw data. The main contribution of this method is the creative calculation of the aromatic carbon rate f_A and the aromatic ring system condensation parameters H_{AU}/C_A from the H-NMR data. f_A is a vital structural parameter that can be correlated with physical properties and conversion performance. H_{AU}/C_A is used to characterize the condensation degree of aromatic structures. To simplify the problem, two hypotheses have been proposed. The first hypothesis disregards the heteroatoms because of their very small percentage in heavy oil. The second hypothesis regards the hydrogen-to-carbon ratio in the saturated portion of the average molecule as the constant 2. The latter hypothesis is crucial, because the carbon distribution information in the average molecular structures could be expressed by hydrogen distribution.

Given an example, Liu et al. [6] used H-NMR to analyze data of the f_A of heavy oil and obtained the following empirical formula:

$$f_A = 1.132 - 0.560(\text{H/C}) \tag{6}$$

Compared with H-NMR data, the use of this formula to calculate the f_A, resulted in a deviation below 5%, which indicates a simple but accurate method to calculate the aromatic carbon rate in heavy oil. In heavy oil molecules, the surrounding aromatic system is often substituted by an alkyl group or naphthenic base. The substitution degree can be described as the aromatic system surrounding hydrogen substitution rate (σ). According to research, the σ in heavy oil average molecular structure is usually around 0.3–0.6.

For some aromatic systems, the aromatic rings are fixed, and the condensation degree of the system can be characterized in terms of aromatic nucleus carbon C_I, fused carbon C_F, aromatic ring number R_A, and naphthenic ring number R_N.

A large amount of information on aromatic and saturated structures can be obtained from NMR and IR. However, differentiation between the naphthenic structures and the alkyl chain through these two methods is difficult. Thus, naphthenic part structure mainly used indirect method to estimate, which was estimated as naphthenic carbon rate f_N, alkyl carbon rate f_P, and average chain length parameter L and so on.

The improved B–L method was successfully utilized [6]. Structures of the residue and bitumen from Daqing, Shengli, Gudao, Renqiu, and some other oil fields were previously analyzed; the following results were obtained:

1. The H/C ratio of vacuum residue in China is roughly 1.6. The sulfur content of crude oil is less than 1%, but its nitrogen content can reach up to 1,000 ppm.
2. The related molecular weight of most residues is approximately 1,000. Therefore, the general molecular formula of $C_{70-100}H_{100-160}S_{0.2-0.7}N_{0.4-1.0}$ may be valid.
3. Compared with the crude oil produced around the world, the metal content in China petroleum shows a low vanadium content but a high nickel content.
4. The resin content in the residue is relatively high at 40–50% but its asphalt content is low.

According to these characteristics, the average values of the structural features of heavy oil can be obtained, and its possible molecular configurations may be evaluated. Combined with quantum chemistry calculations, the most probable configurations of average molecules of heavy oil can be determined. For example, Zhao et al. [7] used the improved B–L method to calculate the HVGO fraction of heavy oil, the average molecular structure of which was calculated to be $C_{24}H_{38}S_{0.33}N_{0.04}$. The most probable conformation of HVGO average molecule was also calculated by the MD and DFT methods according to the stable structure energy minimization principle. From the calculated results, the researchers deduced the reaction process that occurs when the oil molecules come into contact with the zeolite catalyst and the adsorption that transpires on the active sites with benzene.

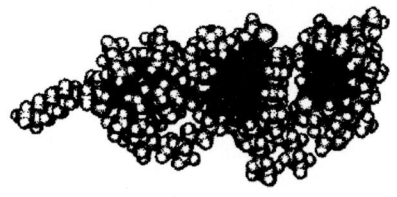

Fig. 1 3D conformation of pitch end-cut molecule with C2 bridge

3.2 Most Probable Molecular Conformation of Heavy Oil

Quantum chemistry was used to calculate the geometrical structure of heavy oil molecules. Pan et al. [8] used the DFT method to calculate the minimum cross-sectional diameter of Daqing heavy oil molecules with the Gaussian 03 program. Their research introduced the concept of molecular dimension, which may be used to predict the minimum cross-sectional diameter of typical heavy oil molecules. This work offered useful theoretical basis for the selection and synthesis of zeolites.

Based on the 2d average molecular unit structure proposed by Zhao et al. [9], Gao and his coworkers [10] used MM and MD simulation methods to determine the lowest energy structures and the most stable 3D conformations for the end cuts of Athabasca bitumen pitch in Canada. Their results provided information on the molecular size and shape. Thus, the 3D conformations of residue molecules were added to a database, including 216 narrow fractions in domestic and overseas locations. Figure 1 shows an example of a 3D conformation of a pitch molecule.

Duan et al. [11, 12] investigated the impact of the side-chain distribution on the 3D molecular structure of heavy gas oil (HGO) from Athabasca and deduced the most probable average molecular conformation. In this work, the optimized steric geometry was determined by MM and MD calculations. The nano-minicrystal structures of the catalyst, as well as the adsorption and hydrogenation reactions on active catalyst surfaces, were simulated via the HyperChem ZINDO/1 semiempirical method from the average model configuration of HGO. A clearer and more intuitive understanding of feedstock molecules and catalysts was achieved.

Zhen [13] and Wang [14] further studied the influence of the distribution of different side chains on the 3D structure and determined the relationship between heavy oil conformations and their physical and chemical properties. The narrow cut of the Daqing vacuum residue was used as a model to calculate the most stable conformation with minimum potential energy by quantum chemistry calculation method, combined with the calculation packages for the HyperChem molecular mechanics (MM+) module and MD simulation methods. Several 3D parameters were simulated according to the parameters that could predict the most extensive average molecular side-chain lengths and distributions.

Based on these data, a reasonable hydrogenation reaction model was proposed by using the MD method for optimization. The reaction path of catalytic hydrogenation was calculated to explain the experimental phenomena observed.

After the successful application to the hydrogenation reaction, quantum chemistry calculations and molecular simulation methods have been extended to some other reactions. For example, the reaction mechanism of catalytic cracking desulfurization of thiophene and the hydrodenitrogenation process of quinoline have been studied by quantum chemistry, and the results show that the calculated reaction pathway of cracking is consistent with the experimental results [13].

The AM1 semiempirical quantum chemical method in the Gaussian 03 program can calculate the resin and supermolecule with multilayered structure and produce better molecular geometries and heats of formation. In addition, intermolecular hydrogen bonds are better described. Thus, this method was applied by Wang et al. [14] to determine the structure of petroleum gums. The optimized geometrical structures of single layer gum (SG), double layer gum (DG), and triple layer gum (TG), as well as their intermolecular interaction energies, were obtained. The structures and properties of the gum were analyzed, and results indicated the planar structures of polycyclic aromatic and alicyclic groups of SG and the stretched structures of alkyl side chains were on the same plane. Polar groups between DG and TG were further revealed to form hydrogen bonds. This phenomenon reveals that the DG and TG structures are difficult to diffuse into the pores of molecular sieves for catalytic reaction because of their supramolecular bulk. In this system, catalytic cracking probably occurs on the surface of the molecular sieve.

Combined with elemental analyses, molecular weight data, and NMR data, the improved B–L method was modified by Ren et al. [15] to construct the average molecular structure of heavy oil fractions. Thus, the heavy oil fraction density was simulated by molecular simulation while the rationality of the presumed average molecular structure was evaluated. The group used Tahe and Suizhong atmospheric residues to construct average molecular structures of saturates, asphaltene, polycyclic aromatic hydrocarbons, and heavy resin. Optimization was completed by the Discovery module in MS software by using MM and MD calculations. Consequently, the global energy minimum conformations and the most stable conformation were obtained. The model compounds and heavy fraction density were also characterized in this work, and the simulated values were clearly consistent with the experimental data. This study indicated that molecular simulations were suitable for calculating the density of heavy oil fractions.

A molecular-level kinetic model has been developed for the pyrolysis of heavy residual oil. Klein and his group [16, 17] modeled the residue structure in terms of three attribute groups: cores, intercore linkages, and side chains.

The Research Institute of Petroleum Processing investigated the hydrogenation process of polyaromatic hydrocarbons (PAHs) [18]. This group constructed a series of PAHs with different aromatic rings to investigate the hydrogenation reaction process from the reaction mechanism via molecular simulation methods. The DMol^3 module in Materials Studio software was used to calculate the reaction heat and energy barrier. The results of this study showed that hydrogenation was affected by neither ring numbers nor the degree of saturation. When an adequate amount of hydrogen radicals is present in the heavy oil system, the radicals can effectively promote the hydrogenation saturation of PAHs and repress the condensation and coking of macro radicals.

Based on actual industrial data, the Research Institute of Petroleum Processing and East China University of Science and Technology used feedstock of deep catalytic cracking (DCC) for molecular simulation by the MC method combined with structural oriented lumping [19, 20]. In this simulation, over 1,000 virtual molecules were constructed, and these molecules showed to be consistent with the composition and characteristics of the feedstock. Results demonstrated that the MC method can simulate the DCC feedstock well on the molecular scale. These materials were further simulated by the MC method according to the characteristics of the delayed coking crude oil [21]. Such studies provide a foundation for constructing molecular reaction dynamics networks, estimating kinetic constants, and simulating product yields.

Zhang et al. [22, 23] worked on heavy cracking feedstock development and built a 10-lumped model. A total of 29 reaction rate constants in this lumped model of Daqing heavy naphtha were obtained by programming the Marquardt++ method in MATLAB. The corresponding pre-exponential factor and activation energy were calculated by using the Arrhenius regression equation. Based on model dynamics analysis, heavy naphtha steam pyrolysis reactions occur as the primary reaction with relatively high regression accuracy. More remarkably, the dynamics parameters were consistent with pyrolysis laws.

High-field Fourier-transform ion cyclotron resonance mass spectrometry (FT-ICR MS) is a very useful physical instrument for analyzing heavy residue compositions. Shi and his group [24] used this method to propose a novel compositional model development methodology and built a structural attribute library for Venezuela petroleum vacuum residue. Here, more detailed molecular composition was determined, including those of polar heteroatom species [25], basic nitrogen compounds [26], and vanadyl compounds [27–29], among others. This method further revealed the dominance and distribution of heavy oil structures.

The true nature of aggregation is difficult to fully understand because of the complexity of the asphaltene composition. However, with the development of computer technology, computational modeling can provide new insights into aggregate structures and the aggregation process. MD, MM, and QM simulations can model the asphaltene aggregation system at the molecular level. One of the most useful aspects of this modeling system is prediction of the detailed structure of the aggregates. Three types of aggregates of asphaltene model compounds have been reported [30, 31]. Several influencing factors, such as the temperature and solvent effects, were also simulated in this work. Another useful application of molecular simulation involved the DFT method for predicting the thermodynamic properties of asphaltene aggregates [32, 33]. The binding energies in different modeling systems were markedly distinct. Besides asphaltene aggregation, the interaction between asphaltene and some other components of heavy oil can also be simulated [34]. The adsorption energy, binding energy, and equilibrium morphology have been calculated and discussed. As previously discussed, several new theories and discoveries are supported by molecular simulations.

4 Applications in Catalyst Development

Catalytic reactions in petroleum processing are mainly based on accelerated reactions which occur at the gas–solid interface. Generally speaking, a gas–solid catalytic reaction has seven steps: [1] diffusion of reactants from the gas phase to the external surface of the catalyst (external diffusion); [2] diffusion of reactants from the external surface to the internal surface of the catalyst (internal diffusion); [3] adsorption of reactants at adsorption sites (adsorption); [4] reactions at active sites (surface reaction); [5] desorption of products from adsorption sites (desorption); [6] diffusion of products from the internal surface of the catalyst to the external surface (internal diffusion), and [7] diffusion of products from the external surface to the gas phase (external diffusion). As each step influences the reaction process and product distribution, catalytic reactions, particularly the adsorption and diffusion reactions, have been extensively studied.

4.1 Adsorption

To maintain good catalytic activity, the adsorption of reactants on the catalyst surface must neither be too strong nor too weak; moderate and fast adsorption are usually necessary. Therefore, investigating the adsorption process is of considerable importance to understand the activity, selectivity, and catalytic mechanism of catalysts. Molecular simulations can shed new light on the findings of the experimental work. Simulations are very useful and accurate methods to estimate adsorption properties, such as the adsorption isotherm, heats of adsorption, and adsorption sites. The adsorption properties of the most widely used zeolites, such as MFI and FAU, have been extensively simulated by using adsorbates varying from long-chain hydrocarbons to bulky aromatic molecules. Overviews of adsorption studies of the linear alkanes and other molecules in MFI zeolite by molecular simulations were separately compiled by Fuchs and Cheetham [35] and Smit and Maesen [36]. In the present paper, we emphasize the adsorption properties of another important zeolite in the petrochemical industry: the FAU-type zeolite.

4.1.1 Adsorption Isotherms

In nearly all simulation studies of adsorption in confined systems, the adsorption isotherm is the first property calculated and compared with experimental results. This isotherm shows the connection between simulations and experiments.

Authenticating Model and Force-Field Parameters Adsorption isotherms play an important role in authenticating models and the applicability of force-field parameters for simulation studies [37]. Verification of a force field is necessary because unsuitable force-field parameters may produce incorrect results when

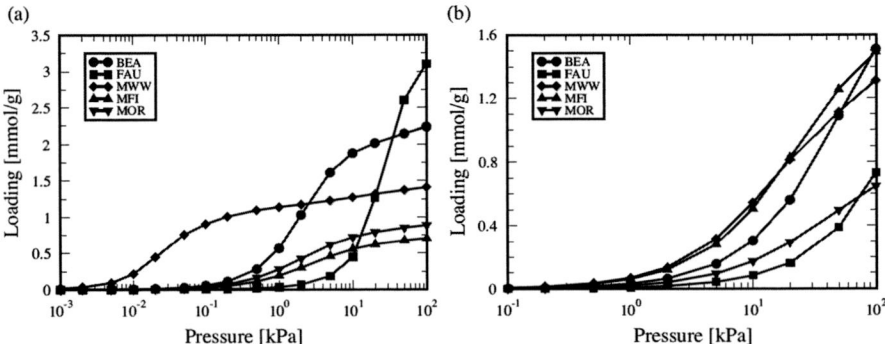

Fig. 2 Simulated adsorption isotherms of (**a**) benzene at 423 K and (**b**) propene at 373 K in zeolite β (BEA), Y (FAU), MCM-22 (MWW), silicalite (MFI), and mordenite (MOR) [49]

applied to different zeolite frameworks. For instance, Calero and other researches [38–45] developed a united atom force field to describe the adsorption properties of linear alkanes in the sodium form of FAU zeolites. The force field was validated by comparing the adsorption isothermals with the corresponding experimental data. Garcia-Pérez et al. [46] extended this force field to the calcium form of FAU zeolites. Granato et al. [47] established a new set of parameters for cation interactions with alkene double-bond carbons and successfully described the equilibrium adsorption isotherms of propane and propylene on 13× zeolite. Based on a proper match between the calculated and experimental adsorption isotherms, the suitable force field and model can be chosen for further calculations. For these comparisons, expression of the adsorption amount in zeolite in terms of the absolute amount per volume of adsorbent is more suitable than the conventional adsorbed amount in moles per gram of adsorbent because the adsorption columns are usually designed in terms of the bed size rather than the mass of the adsorbent. The adsorbed amount can be transformed into moles per volume of adsorbent by using the grain density. Most experiments rely on measurement of the weight increase of a zeolite sample; thus, pure-component adsorption isotherms that are directly related to the number of adsorbed molecules are significantly easier to measure than the isotherms of mixtures [38].

Molecular Interactions: Lower Pressure Many studies have taken advantage of the characteristics and differences of isotherms to understand the interaction between adsorbates and zeolite and explain the adsorption behaviors of various adsorbates in different frameworks [48]. Figure 2 shows the adsorption isotherms of benzene and propene in various all-silica frameworks [49]. At the temperatures and pressures in this study, benzene adsorption begins at much lower pressures than propene adsorption because the attractive interactions of benzene with the framework are much stronger. Furthermore, the slope of the MWW adsorption isotherm at low loadings (Henry coefficient) is large because the size of benzene is commensurate to the size of the cavity in MWW.

Molecular Interactions: Saturation Pressure The saturation adsorption amount can be estimated from the adsorption isotherms as long as the simulations are conducted at sufficiently high fugacity [50]. The amount of adsorption at saturation is also determined by the molecule–molecule and molecule–framework interactions [51], which are related to the size and structure of the molecules, dimension and composition of the framework, and packing pattern of the molecules in pores. Zheng et al. [52] performed grand canonical Monte Carlo (GCMC) simulations to study the adsorption of propane in FAU-type zeolites X and Y with varying contents of the exchange cations Na^+/Ca^{2+}. For a given ratio of Si:Al, the saturate adsorption amount of propane in both sodium and calcium FAU increases with increasing number of cations because of the production of extra adsorption sites, which leads to a higher amount of adsorption. For a given non-framework cation number density, higher ratios of Si:Al promote higher adsorption amounts caused by reductions in charge density, i.e., fewer calcium cations and more sodium cations make the interaction energy more negative and consequently increase the affinity between propane and non-framework cations.

Molecular Interactions: Selectivity Based on the interactions of target guests in mixtures, adsorption isotherms can also be used to characterize adsorption selectivity. Ban et al. [49] simulated the adsorption isotherms of benzene and propene in equimolar (50/50) mixtures and obtained high selectivity toward benzene under various conditions in four zeolites: MOR, beta, Y, and MCM-22. These results may be explained by the fact that when the size of the micropores is close to the kinetic diameter of the target guests, the interaction energy of the guest molecules is lower than that of other adsorbates and the target guests are preferentially adsorbed. Given that the high adsorption selectivity of benzene is guaranteed by the micropore sizes, which is close to the kinetic diameter of benzene regardless of the micropores topology, potential zeolite catalysts for benzene alkylation must feature micropores with sizes similar to benzene.

Flexibility Although the most widely used simulation method to calculate the adsorption isothermal is standard GCMC simulation, the limitation of constant volume during adsorption makes it incapable of modeling framework flexibility, as determined by many experimental studies [53]. To optimize accuracy and efficiency, the constant-pressure ensemble (the osmotic ensemble) [54], which allows unit-cell expansion (and contraction) upon adsorption, can be adopted to model such flexibility. For instance, Santander et al. [55] compared pure MC sampling with hybrid MC-MD simulations and found that hybrid MC-MD sampling of the osmotic ensemble and MC sampling both reproduce the Henry's law regime; however, the saturation loadings of these systems by hybrid MC-MD sampling are more extensively than that of standard MC sampling because the MC-MD approach naturally allows for locally anisotropic volume changes wherein some pores expand.

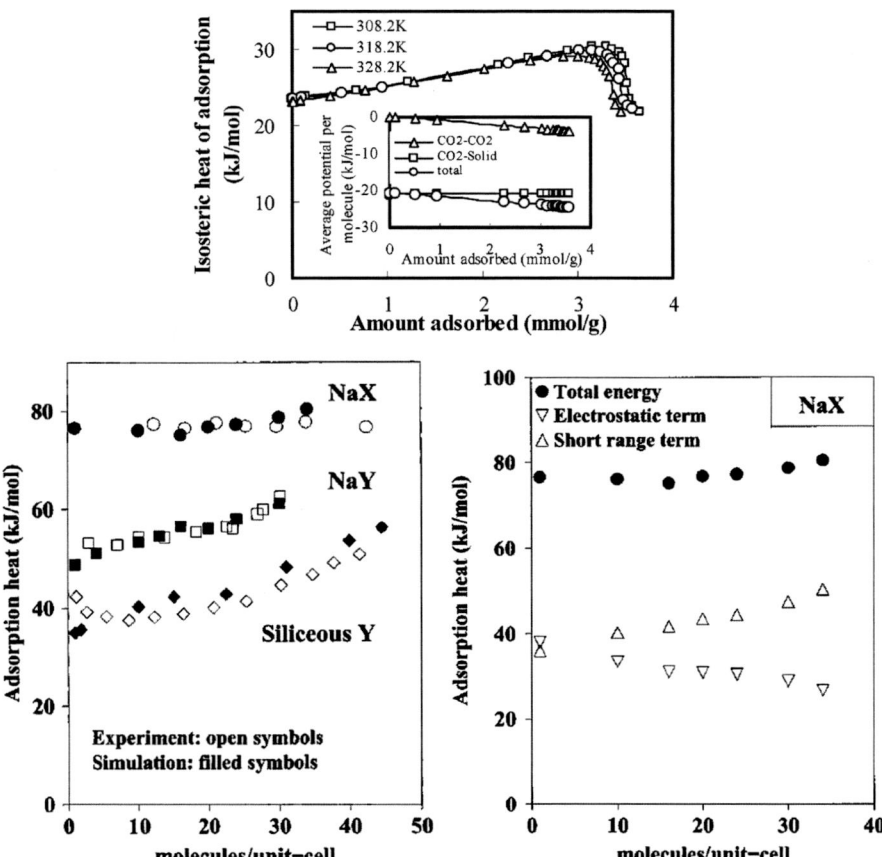

Fig. 3 (**a**) Isosteric heat of adsorption for CO_2 on silicalite at 308.2, 318.2, and 328.2 K. (**b**) Observed and calculated heats of adsorption of chloroform at room temperature in the three zeolite hosts siliceous faujasite, NaY, and NaX. (**c**) Calculated heats of adsorption for siliceous NaX decomposed into short-range and long-range contributions

4.1.2 Adsorption Thermodynamics

Similar to adsorption isotherms, adsorption energy, which is often described in terms of isosteric heat of adsorption Q_{st}, is another important quantity that can be measured both in experiments [56–58] and calculations [59]; thus, this parameter is often used to investigate the quality of the chosen force field and model. Figure 3a shows the Q_{st} of adsorption for CO_2 as a function of adsorbed amount in silicalite at different temperatures. The Q_{st} of adsorption under zero loading, at around 23.5 kJ/mol, is independent of temperature. This value is comparable with the experimental data of Golden and Sircar [58, 60] (24.07 kJ/mol), Sun et al. [61] (28.5 kJ/mol), Dunne et al. [58] (27.2 kJ/mol), and Choudhary and Mayadevi [62] (20.0 kJ/mol).

For the considered loading range, Q_{st} shows similar trends despite loading at different temperatures; this trend is found in many other guest/host systems [63].

In some adsorbate molecules, changes in Q_{st} under different loadings are debated. For example, Q_{st} decreases with the loading in benzene. Barrer et al. [64] found a slight decrease in molar differential heat of adsorption from 18 kJ/mol to 17 kJ/mol with increasing coverage for dehydrated zeolite X (Si:Al = 1.33) through adsorption isotherms. A similar result has been reported for the adsorption of benzene onto zeolite Na–X [63]. This result is usually explained by adsorption on less favorable sites, which decreases the sorbate–zeolite interaction at higher loading. In addition, with a general trend toward higher heats versus coverage, several minima in the curve of Q_{st} were found during adsorption of benzene onto silicalite at 301 K [64, 65]. The authors attributed this phenomenon to the mutual interaction between benzene molecules forming dimers in one stage and larger clusters in another. Q_{st} has been observed to generally increase with increasing loading. This result has been reported in many studies. Thamm and Regent [66, 67] described increasing Q_{st} with coverage from 18.6 to 20.8 kJ/mol during benzene adsorption onto zeolite Na–Y (Si:Al = 2.37) at 301 K and attributed the high heats of adsorption observed at higher loadings to favorable sorbate–sorbate interactions. The adsorption heat slightly increases with the adsorption loading and eventually rapidly decreases [68–71]. Still take Fig. 3a as an example; the average potential energy per CO_2 molecule on silicalite at 318.2 K is split into the adsorbate–adsorbate and adsorbate–adsorbent interactions. The increase in Q_{st} with increase loading can be attributed to enhanced interactions between CO_2 molecules. The interaction between CO_2 molecules and the adsorbent remains essentially unchanged, which signifies that CO_2 adsorption sites in this solid material are relatively homogeneous. The presence of maximum values in Q_{st} at high adsorption occupancy is due to the increasing importance of the bulk fluid term in Q_{st}, which is calculated by using the Peng–Robinson equation type of zeolite; different types of loading dependence of Q_{st} have been determined by both simulations and experiments. For instance, the loading dependence of Q_{st} sorption on silicalite was found to differ between aromatic molecules (benzene, toluene, and ethylbenzene) and n-hexane; here, the state of p-xylene on silicalite is regarded as the intermediate between these two cases [72]. This result is directly attributed to the different change trends of sorbate–sorbate interactions in polar and nonpolar sorbates.

Several authors have reported that the Si:Al ratio and the type of cations can also significantly affect the adsorption heat because of enhanced interactions between the adsorbate and the zeolite framework [73–76]. Figure 3b shows the simulated and experimental calorimetric heats of adsorption of chloroform at room temperature in three FAU zeolites (i.e., siliceous faujasite, NaY, and NaX) with different Si:Al ratios. The simulations are remarkably similar to the experimental findings. Increases in adsorption energy over the whole loading range follow the order siliceous faujasite < NaY < NaX. The slight decrease in energy observed at low loadings in siliceous faujasite is probably caused by a low concentration of silanol defects. This result may be explained by decomposing the total host–guest interaction energies into their short-range and long-range components. Only NaX data are shown in Fig. 3c. Similar short-range interactions are observed in all three hosts, and the higher affinity of chloroform for more polar zeolites clearly stems from stronger electrostatic interactions occurring in these cation-containing systems. In contrast to the significant increases in adsorption energies observed upon loading in siliceous faujasite and NaY, the chloroform/NaX system shows a relatively flat profile. This profile may be explained by the contribution of SIII cations in the case of zeolite NaX, where cancellation between short-range and electrostatic interactions leads to a heat of adsorption that is invariant with the coverage, as shown in Fig. 3c.

Further explanations based on the simulation of relative contribution of adsorbate–adsorbate, adsorbate–adsorbent, short-range, and electrostatic interactions are necessary to understand the changing trend of Q_{st}.

4.1.3 Adsorption Sites

The adsorption of guest molecules in zeolites, which have micropores of molecular dimension, is attributed to local effects rather than the properties of the whole solid [77]. Apparently, molecules adsorbed on isolated specific sites can also moderately extend interactions with neighboring sites and "feel" the potential of large parts of the solid depending on the shape of the surface [78]. Therefore, the practical catalytic and sorption properties of zeolites may be determined by the number, strength, distribution, and accessibility of the catalytic sites [79, 80]. For example, during petroleum cracking on HY zeolites, low Si:Al ratios (predominantly weak Brønsted acid sites) lead to extensive hydrogen transfer, resulting in high gasoline yields and conversion levels with low olefin yields and relatively high coke formation. Conversely, for zeolites with relative high Si:Al ratio, i.e., few but predominantly strong Brønsted acid sites, the opposite is true. Predictions of energetics and the site locations of sorbates from MC simulations are expected to provide pertinent information in this field. A large number of groups have computed the adsorption sites of many adsorbates in various zeolites [81–83]. In this paper, we focus on the molecular simulation studies of the adsorption sites of aromatic species in FAU zeolite, which is the main component of FCC catalysts.

Si:Al: Silica-FAU The type and energy of adsorption sites are directly related to the Si:Al ratio of FAU zeolite. A purely siliceous faujasite structure is represented by the formula $(SiO_2)_{192}$ and has an electrically neutral framework. Based on the simulation of Brémard et al. [84] in siliceous FAU, $Mo(CO)_6$ and C_6H_6 molecules are randomly distributed in the void space rather than located at specific sites. However, Rungsirisakun et al. [85] found that benzene molecules prefer to be adsorbed parallel to the surface of the sodalite cage above the six-membered ring. The presence of aluminum(III) introduces charge defects in the zeolite framework, which must be compensated with some non-framework counterions to achieve global electroneutrality. The number of these ions required per zeolite unit cell depends on the aluminum content and their own electric charge.

Si:Al: NaY For the alkali extra-framework cations in faujasite, four crystallographic sites defined as: sites SI (located in the center of the hexagonal prism), SI' (present in the sodalite cage toward the hexagonal prism), SII (located in the 6-ring windows of the supercage), SIII (in the 12-ring windows of the supercage), and sites III' (in the 12-ring windows of the supercage) [86, 87]. Sites of FAU zeolites substituted with monovalent cations M^+ such as Na^+ were most widely studied [78]. The first adsorption simulations considering benzene/NaY system [88] identified two adsorption sites with approximately half the molecules occupying a position in the vicinity of the SII site with an average energy of -75 kJ/mol and the remaining molecules sitting in the site denoted as W involve benzene in the 12-ring window with a mean energy of -60 kJ/mol at 326 K. Neutron diffraction studies [83, 89] and theoretical studies [84, 88, 90] have obtained the same two binding sites for benzene in NaY at low temperatures.

Si:Al: HY Unlike NaY, the adsorption sites of HY zeolite can be related not only with one compensated protons (proton site). That is to say, for sufficient adsorbates, the distance between two protons can be adequately close to form the "supersite," which involves more than one proton atoms. Jousse et al. [91] performed molecular docking to study the location of benzene in a model zeolite HY (Si:Al = 2.43). Multiple adsorption sites in accordance with infrared measurements were detected including the "supersite." Jirapongphan et al. [92] investigated the adsorption of benzene in HY zeolite by a newly developed supercage-based molecular docking simulation. Six different adsorption sites of benzene in HY were reported at the loading of 4 mol/UC, including a new stable adsorption site in addition to confirming the multiple benzene adsorption sites as Jousse reported [91].

Si:Al: Our Work We [93] performed MC simulations to study the adsorption site of benzene in HY zeolite with different Si:Al ratios, focusing on finding all of the possible supersites and understanding the influence of Si:Al ratio on the reaction mechanics. Eleven types of adsorption sites were observed, including five reported sites [i.e., H1, H2, U4, U4(H1), and W] and six newly found sites [i.e., W(2H1), U4 (2H1), H1(H2), U4(H1,H1), H1(H2,H1), and U4(H1,H1,H1)], which were "supersite" with more than one proton. The stability of the sites found in the 28Al model showed the order U4(H1,H1,H1) > U4(H1) > H1(H2,H1) > W (2H1) > U4(H1,H1) > H1(H2) > H1 > H2 > U4 > U4(2H1) > W. Increasing the number of zeolite protons resulted in increases in the proportion of supersites, which, in turn, enhanced the adsorption energies of the sites. For HY zeolite models containing different numbers of protons with the same H1:H2 ratio, the amount of the most stable adsorption sites containing H1 proton increased, whereas the amount of the most stable adsorption sites containing H2 decreased, with increasing number of protons.

Loading: Reported Studies The ratio of the population of the sites depends on the loading of the adsorbate. In general, at high loadings, confinement of the molecules

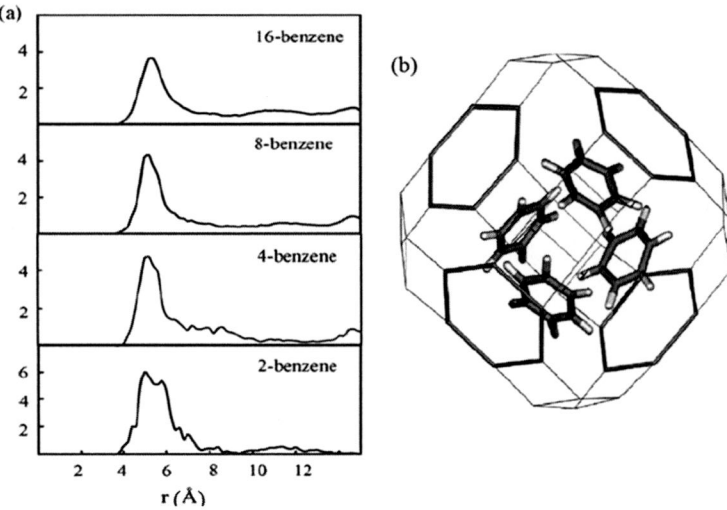

Fig. 4 (a) Radial distribution functions of benzene in siliceous FAU at different loadings. (b) A snapshot of four benzene molecules in one supercage

to porous voids constrains the molecules to reside in energetically unfavorable sites through intermolecular contact [84]. Molecular dynamics simulations of benzene in siliceous FAU have previously been performed at different loadings (2–16 molecules per supercell) [85]. Pair correlation functions, which represent the conditional probability of finding two particles at a given separation, normalized to unity at long distances, have been used to identify the adsorption sites at different loadings. The first peak corresponds to a contact benzene dimer that has its maximum at about 5.2 Å (Fig. 4a) for all four loadings of FAU and refers to benzene located on six-ring windows. Figure 4a shows that the height of the peak maximum in the resource description framework decreases with the loading and shifts toward larger distances, thereby indicating increasing disorder in the adsorption sites. By investigating the snapshot of a typical configuration of four benzene molecules in one supercage of FAU, a range of distorted V-shaped structures is observed, as shown in Fig. 4b. Results on the loading of 16 molecules/UC are inadequate to confirm such a conclusion, however, and calculations on saturated loadings must be performed to examine possible steric effects under higher molecular loadings.

Loading: Our Work We have calculated the adsorption of benzene at the full loading range (1–44 mol/UC) in four HY zeolite models with different Si:Al ratios (∞, 12.71, 5.86, and 2.43). Two adsorption mechanisms were found by investigating the adsorption sites of benzene. Below the inflection point (up to 32 molecules/ UC for all Si:Al ratios), benzenes are located on the supercage site with an ideal adsorption geometry configuration, in accordance with the previous studies. Above the inflection point, benzene molecules tend to insert into the space between existing adsorbed benzenes, and no notable rearrangement is observed for previously adsorbed benzenes (the insertion mechanism). The proposed adsorption mechanism is independent of the Si:Al ratio, and the inflection point shifts toward higher loadings for zeolite with lower Si:Al ratios.

Accessibility The accessibility of acid sites also plays an important role and may explain, in some cases, the differences observed in the separation properties of zeolites. For instance, the preferential adsorption of NaY for *m*-xylene compared with *p*-xylene and the exchange of Na^+ ions with K^+ ions leads to a reversal in the adsorption selectivity found in many experimental studies [94]. Lachet et al. [72] explained this by detailed analysis of the adsorption sites, which revealed extremely different molecular mechanism of the cage filling in zeolites NaY and KY toward xylene molecules. Regardless of the loading, a unique adsorption site is revealed in the supercage of the zeolite NaY,; this site is located in front of sodium cations in site II with xylene/NaY interactions stronger for *m*-xylene than for *p*-xylene. In KY, three adsorption sites of different energies were observed. The strongest xylene/KY interactions have been observed only for the so-called II/III sites in the case of *m*-xylene and for both II/III and II/W sites in the case of *p*-xylene. Considering steric effects, sites II/III cannot accommodate more than two molecules per supercage, whereas sites II/III and II/W can altogether accommodate up to three molecules per supercage. As a consequence, adsorption of a third molecule is unfavorable to *m*-xylene. Another simulation study [95] on different adsorbates in HY zeolite also confirmed that the adsorption site in FAU zeolite varies for different adsorbates.

Molecular simulations can provide us detailed information of adsorption properties at the molecular level; in such simulations, adsorption isotherms, heats of adsorption, and Henry coefficients can be compared with the experimental data to validate the rationality of the methodology and model. To date, systematic studies on how zeolite Si:Al ratios, cation charges, and loading of the adsorbate affect adsorption are fairly limited. Fundamental knowledge is necessary to understand the adsorption mechanism fully. Binary or multicomponent mixtures have not received as much attention as pure sorbates, although these mixtures are of most interest in practical applications. Only a small number of recent studies are available, and the majority of these studies focus mainly on sorption thermodynamics.

4.2 Diffusion

The diffusivity of residue fractions through membranes is often investigated by using a diaphragm cell. Results show that the size of molecules and pores is a crucial factor affecting the diffusion transport behavior of residues through membranes. In the composition and configuration of molecules, it may also affect hindered diffusion of residue molecules. The hydrodynamic size of residues has also been determined through membrane diffusion experiments [96]. Notable polydispersities for residue fractions have been observed from the variation of diffusion

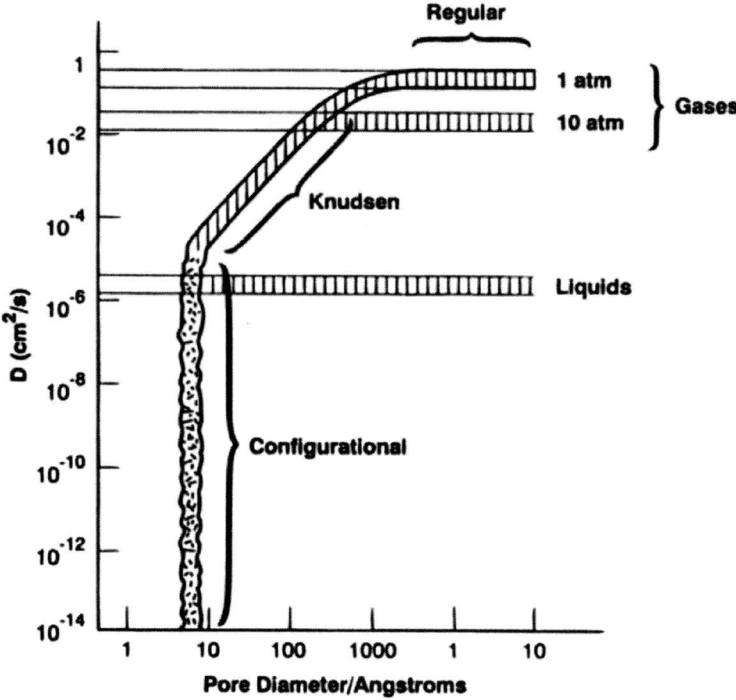

Fig. 5 Relationship between diffusivity and aperture size pore in regular diffusion, Knudsen diffusion, and configurational diffusion [99]

coefficients. Further studies reveal that aggregation of maltene and asphaltene molecules can occur at low concentrations [97].

Weisz [98] summarized the diffusion of gaseous molecules in a porous solid as conventional diffusion (regular regime), Knudsen diffusion (Knudsen regime), or configuration diffusion (configurational regime). Conventional diffusion refers to the diffusion in macropores when the pore size is greater than the mean free path of gas molecules, and the diffusion coefficient is equivalent to the ordinary gas diffusion coefficient calculated by reductions in porosity. Knudsen diffusion refers to the diffusion in mesopores when the pore size is shorter than the mean free path of gas molecules, an example of which is the diffusion in silica–alumina catalysts. Zeolite crystals have strict, regular, and precise microporous structures, and the molecule size usually approaches the size of the zeolite pores; therefore, diffusion in zeolites may be referred to as configurational diffusion. The relationships between diffusivity and aperture size are shown in Fig. 5.

To design and develop better separation and catalytic process technologies, a proper quantitative description of the diffusion of guest molecules within meso- and microporous structures (Knudsen and configurational regimes) is necessary [100]. Diffusion in zeolites and carbon nanotubes has been widely investigated. Molecular simulations are important adjuncts to experiments and can explain and describe a variety of experimental data and observations. A detailed overview of Krishna's work demonstrated that the fundamental understanding of diffusion of guest molecules in porous structures is significantly aided and enhanced by the use of molecular simulation techniques. In this paper, we focus on simulation studies of diffusion in zeolites (configurational regime) and hierarchical zeolites (Knudsen regime).

4.2.1 Diffusion in Zeolites

T Loading The self-diffusion of many adsorbate molecules in various zeolites has been studied by molecular dynamic simulations at various temperatures and loadings. Self-diffusion coefficients generally increase with increasing temperature because the increasing kinetic energy of the gas molecules enlarges their mean free path [101].

Both diffusivity and activation energy for adsorbate diffusion are loading dependent [102]. However, for a given structure, prediction of whether the diffusion will decrease, increase, or remain constant when the loading is increased is usually impossible [103]. Transition-state theory simulations can be used to gain insights into the diffusion properties of zeolites, as studies on this topic are often confined to infinite dilution limits [104]. Molecular dynamics simulations can easily obtain the loading dependence of diffusivities [105]. However, similar to experimental results, diffusion coefficients as a function of loading determined through MD simulations are also difficult to rationalize at the molecular level.

Confinement As inherent characteristics of different diffusion regimes, besides loading and temperature, an important factor influencing the magnitude of diffusivity is the degree of confinement [100]. Xiao and Wei [106] reported a relationship between diffusivity and the ratio of the kinetic diameter to the channel diameter (r/d): The diffusivity is more or less independent of this ratio if r/d is smaller than 0.6, which corresponds to diffusion taking place in the Knudsen regime. If the kinetic diameter approaches the channel diameter, diffusivity sharply decreases, a phenomenon known as configurational diffusion. This decrease is confirmed in Fig. 6 [107], which shows that surface diffusivity is nearly constant at kinetic diameters of up to about 0.3 nm, corresponding with $r/d = 0.55$. Above this ratio, the corrected diffusivities decrease strongly, on the half logarithmic scale, with the kinetic diameter. This decrease indicates that diffusion occurs in the Knudsen regime only for helium, neon, and hydrogen; for other gases that are of more importance in the industry, diffusion occurs in the configurational regime. As the size of the adsorbate further increases, diffusivity is generally characterized by the levitation (window) effect.

Adsorbate Size: Levitation (Window) Effect An intriguing "window effect" associated with mass transport in zeolites was first reported through a gravimetric study by Gorring for normal paraffins in Zeolite T. Gorring also noted that the observed dependence of diffusivity on the chain length can be used to justify the

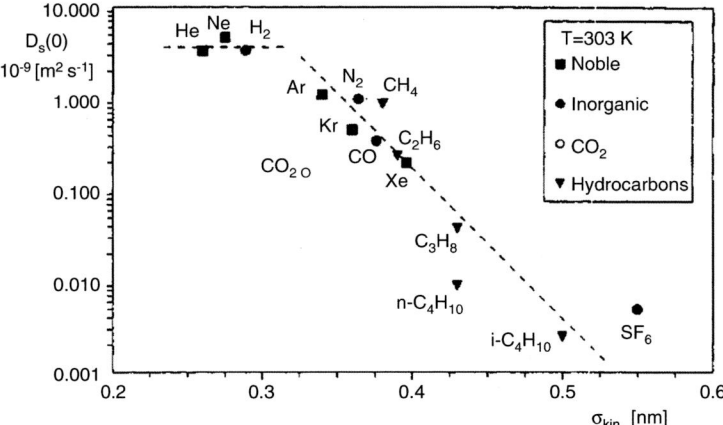

Fig. 6 Diffusion coefficient as a function of the kinetic diameter. The *dashed lines* are meant to guide the reader's eye [107]

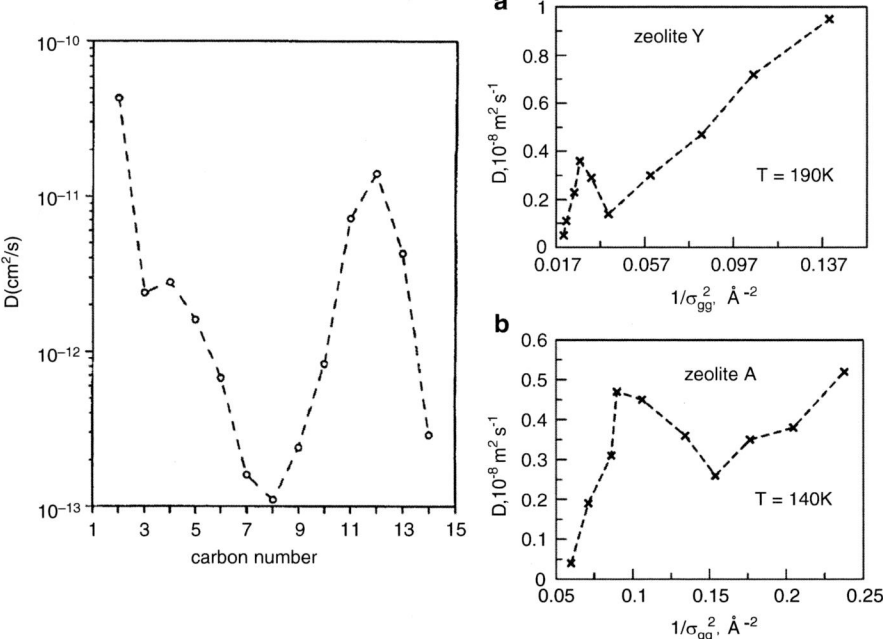

Fig. 7 (**a**) Diffusivities of *n*-paraffins in zeolite T at 300°C. (**b**) Variation of self-diffusivity D as a function of the reciprocal of the square of the guest diameter for model guests in zeolites Y and A interacting via the Lennard–Jones potential [113]

unexpected product distribution obtained from the catalytic cracking of n-C_{23} over H-erionite reported by Chen et al. The diffusion coefficients measured by Gorring (shown in Fig. 7a) demonstrate an unexpected maximum value for dodecane following the existence of a minimum for the diffusivity of n-octane, the difference between the two extrema being of two orders of magnitude. Gorring's data were confirmed by Magalhães et al. [108] by using gravimetric sorption/desorption techniques. The effect observed leads to a maximum value of D, which is also known as the levitation effect [109]. No evidence of a window effect has been suggested for the diffusion of linear paraffins in zeolites 5A, silicalite, and NaX [110]. However, simulation studies suggest that the size-dependent diffusivity maximum exists in a wide class of porous solid system regardless of the geometrical and topological characteristics of the pore network provided by the porous solid, as shown in Fig. 7b for Y and A zeolites [111]. As suggested by previous molecular dynamic investigations, the size dependence of self-diffusion in guest-porous solids may originate from the mutual cancellation of forces that occurs when the size of the diffusant is comparable with the size of the void [112]. The diffusivity maximum is highly pronounced in the presence of electrostatic interactions. Disorder appears to only slightly reduce the height of the diffusivity maximum and does not eliminate this parameter completely.

4.2.2 Zeolite Modification

Number and Type of Framework Cations The diffusion of adsorbates inside a zeolite framework can be moderately hindered by larger or increasing numbers of cations contributing additional electrostatic forces and steric effects. As in the case of n-octane [114], diffusion coefficients are larger in silicalite than in Na-ZSM-5. This result is in agreement with quasi-elastic neutron scattering [110] findings for longer n-alkanes, which reveal that the ratio of diffusivities in silicalite compared with those in Na-ZSM-5 appears to increase with increasing chain length. Details of diffusion in hierarchical zeolites are presented in Sect. 4.2.3.

4.2.3 Diffusion in Hierarchical Zeolite

To overcome the confinement effect in zeolites, which is caused by the sole presence of micropores rendering low utilization of the zeolite active volume in catalyzed reactions, the exciting field of hierarchical zeolites has rapidly evolved. In hierarchical zeolites a single material couples the catalytic power of micropores with the facilitated access and improved transport inherent a complementary mesopore network. A critical review of Pérez-Ramírez [115] focused on hierarchical zeolites to illustrate the enhanced utilization of microporous crystals in catalysis by advances in material design. The concrete examples are FCC catalysts with hierarchical structures in use in industry. The FCC catalyst is composed of a principally USY zeolite with strong acidity and mesoporosity generated by steam

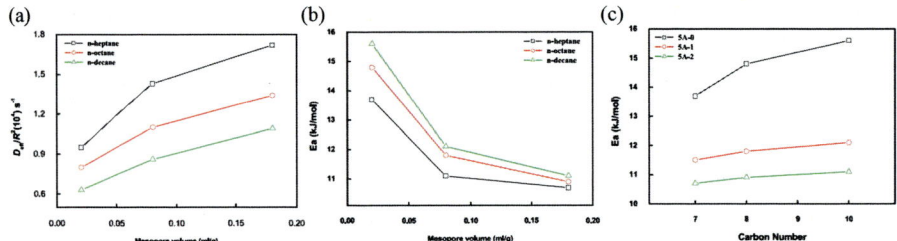

Fig. 8 (a) Effects of the mesoporous volume of zeolites on the effective diffusion time constant at 333 K. (b) Changes in activation energy E_a as a function of the mesoporous volume of zeolites. (c) Changes in the E_a of different 5A zeolites as a function of carbon number [118]

dealumination treatment, also mixed artificially with a macroporous matrix, usually amorphous silica, alumina, or silica/alumina with clay.

Experimental Studies Experimental studies discussing the kinetics and diffusion of hydrocarbons in hierarchical zeolites began to grow only in the last few decades [116]. Figure 8 shows that introduction of mesoporous structures exerts strong influences on the overall transport rate of linear paraffins; in fact, diffusivity increases with the mesoporosity of 5A, which is consistent with an earlier report on NaCaA [117]. The strong relationship between micro-/meso-structures and diffusivity in zeolites may be confirmed by these results.

Simulation Molecular simulations have been widely used to successfully predict the diffusion coefficients of ethane and propene in various zeolites, providing results that compare well with the experimental data as long as the correct combination of framework/force field is chosen [119]. Some simulation studies considering hierarchical materials, including nanostructured porous carbons, ordered mesoporous silicas such as MCM-41 [120] and SBA-15 [121], and hierarchical zeolites [122], have been conducted. Bhattacharya [116] developed the methodology to provide a procedure that considers both large cylindrical mesopores and smaller micropores in the pore walls. The methodology for modeling the SBA-15 structure involves molecular and mesoscale simulations combined with geometrical interpolation techniques. They observed that the presence of micropores leads to increased adsorption at low pressures compared with the case of a model without micropores in the pore walls. The capillary condensation that appears at higher pressures is mainly controlled by the mesopore diameter and not influenced by the presence of micropores. Chae [123] conducted a detailed study of the hierarchical nanostructured porous carbons within the framework of MD simulations based on the reactive state summation force field; the results of this study revealed that the template model can successfully reproduce hierarchical porous carbon structures.

FAU: Reported To the best of our knowledge, only one atomistic simulation study on diffusion properties in hierarchical FAU zeolites is available. Coasne et al. [124] investigated the adsorption and transport of a small adsorbate (nitrogen) in hierarchical FAU models by using MC and MD simulations and observed fast transport in mesopores and slower diffusion in the micropores caused by increases in loading. The group also found that flow in hierarchical materials is larger than that in a single mesopore, which is attributed to the transfer between micropores and mesopores. The diffusion of more complex adsorbate-like benzene, which is of greater concern in the petrochemical industry, is not considered in this literature.

FAU: Our Work We [125] previously studied the adsorption of aromatic molecules (i.e., toluene, styrene, o-xylene, m-xylene, p-xylene, 1,3,5-trimethylbenzene, and naphthalene) in all-silica FAU zeolite. In the case of monoaromatics, two-stage "ideal adsorption" and "insertion adsorption" mechanisms were investigated by careful inspection of the locations and distributions of the adsorbed toluene molecules. The number of C atoms per unit cell, which corresponds to the inflection point of adsorbate loading (CI–P), was defined as a valid and convenient characterizing factor describing the packing efficiency of monoaromatics in the FAU zeolite. A three-stage mechanism was also proposed, for naphthalene, a type of diaromatic, which contains the first two stages, especially, a third stage of "over-ideal adsorption." The so-called over-ideal adsorption is considered as such because the naphthalene molecules initially occupy S sites non-ideally at loadings that approach saturation, leading to more localized feature in the adsorbates. Details of the relevant adsorption mechanism can be used to explain the loading dependence of isosteric adsorption heat on the aromatics concerned.

As previous results are generally in accordance with the expectation that hierarchical zeolites benefit more from extensive diffusivity and site accessibility than their monomodal counterparts, the relevant mechanism has been used in commercial catalysts [126–130]. However, the diffusion characteristics in microporous channels arising from introduction of mesoporosity remain unclear. Molecular simulations can provide further understanding of such problems.

4.3 Catalyst Structure

As previously described, the adsorptive and catalytic properties of zeolites are determined by the location and strengths of acid sites, which are decided by the Al atom content and their distribution in the zeolitic framework [131]. Therefore, investigations of the relationship between distribution of Al and activity of zeolite are significant. However, Si and Al atoms cannot be distinguished by NMR or any other technique [132], and the exact location of Al in the zeolite framework cannot be determined by experimental methods [133]. The merits of DFT render it an interesting technique for research in this field. In the present paper, we focus on the relationship between the number and distribution of Al atoms and the activity of a Brønsted site in acid zeolites. During DFT studies, the activity of a Brønsted acid site can be measured in terms of deprotonation energies [134], OH-stretching vibrations [135], adsorption energies of the base molecules [136], and O–H lengths [137].

Table 1 Acid strengths of T sites at different molar ratios [133]

Molar ratio	OH	E_{DEP} (kcal/mol)
47	3,621.4	0
23	3,557.3	7.78
15	3,548.2	13.7
11	3,643.2	6.05
8.6	3,606.9	6.69
7	3,608.1	6.8
5.8	3,607.9	8.92

Strength of the Brønsted Acid Site: Al Concentration The acid strength of zeolite is related to its Al concentration, the distance between Al atoms, the different Al sites, and the species and number of trivalent substituent atoms. Sierka et al. [138] discovered that the Brønsted acid strength of Y zeolite is determined by the number of nearest neighboring Al atoms in the Al–O(H)–Si bridge through the Hartree–Fock and density functional (B3LYP) methods. This group found that clusters with larger numbers of Al consistently produce larger deprotonation energies. While low Al concentrations can create strong acidic sites in a lattice, a lattice with high Al concentrations shows weaker acidic sites with increasing Al. Pine et al. [139] obtained similar findings by using an experimental program and data on zeolite stability. The group then proposed next-nearest-neighbor (NNN) theory stating that acidic strength decreases as the number of Al atoms occupying NNN sites increases. Wang et al. [133] calculated the energy of deprotonation and OH-stretching vibrations in a Y zeolite model with different Si:Al ratios, as shown in Table 1. By comparing data, they found that acid strength increases with increasing Si:Al ratio.

Strength of the Brønsted Acid Site: Distance Between Al Atoms and Different Al Sites Zhou et al. [140] researched Brønsted acidity in supercages of MCM-22 with double Al substitution at NNN and next-next-nearest-neighbor (NNNN) sites by the DFT method and found that zeolite acidity increases as the distance between the two Al atoms increases. The acidity achieved at the NNNN position is stronger than that achieved at the NNN position. This result was also confirmed by Wang [133], who observed when two Al atoms lie in NNN position relative to each other; the resulting acid strength is weak. Different Al sites exert important influences on the activity of zeolite. Zhou et al. [141] studied the acid strength of different protonic acid centers in MCM-22 zeolite (Fig. 9) by calculating their interaction energy. In their work, ammonia, pyridine, nitrogen, and ethane molecules were selected as probes, as shown in Table 2. All of the adsorption energies obtained reveal that $Al1(O_3H)Si_4$ and $Si1(O_3H)Al_4$ sites display similar acidities, whereas $Si_2(O_3H)Al_3$ exhibits slightly lower acidity than the other. These results reveal that the strength of protonic acid in supercages is relatively stronger than that in 10-MR sinusoidal systems, which is in good agreement with experimental findings. Adsorptions on $Si_2(O_{11}H)Al_3$ and $Al_2(O_{11}H)Si_3$ are equivalent, which means adsorption is more affected by the zeolite structure than by its chemical composition. Zheng [142] reported that accessible Brønsted acidic protons most likely

Fig. 9 Location of B-acid sites in MCM-22 zeolite

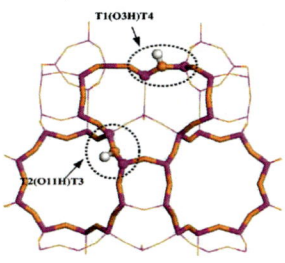

Table 2 Calculated interaction energies between the adsorbing bases and the acidic center based on 8T cluster model (KJ/mol) [141]

	NH$_3$	C$_5$H$_5$N	C$_2$H$_4$	N$_2$
Al$_1$–O$_3$H–Si$_4$	122.82	111.77	102.66	15.52
Si$_1$–O$_3$H–Al$_4$	126.23	110.69	101.8	15.18
Si$_2$–O$_3$H–Al$_3$	104.6	79.26	87.98	10.5
Al$_2$–O$_{11}$H–Si$_3$	111.56	81.52	89.79	11.2

reside in supercages and found that the Al1-OH–Si$_2$ site in MCM-22 zeolite exerts the highest acid strength (as probed by TMPO) through DFT calculations.

Strength of the Brønsted Acid Site: Species and Number of Trivalent Substituent Atoms The acidity of zeolites is provided by the proton, which balances the anion on the framework; thus, the chemical composition of the framework, such as the species and number of trivalent substituent atoms, exerts an important influence on the acidic properties of zeolite. Previous studies focused on the acidity of isomorphously substituted zeolites. Chu and Chang [143] reported that acid strength increases according to the order of B-ZSM-5 < Fe-ZSM-5 < Ga-ZSM-5 < Al-ZSM-5 by using IR spectroscopy and temperature-programmed desorption–NH$_3$ (TPD–NH$_3$) methods. However, the acidity of Brønsted sites in zeolite is difficult to observe in actual experiments. Therefore, theoretical studies, which can provide information at the atom level, are necessary. Wang et al. [144] studied the relative acidity of T-MCM-22 (T = B, Al, Ga, and Fe), including its proton affinity, bond length, and bond angle, OH-stretching frequencies, and charge on the acidic proton by the DFT method. By analyzing the data obtained, they concluded that the acidity of T-MCM-22 zeolites increases according to the sequence B-MCM-22 < Fe-MCM-22 < Ga-MCM-22 < Al-MCM-22, which is in good agreement with the experiment and research of Chatterjee [145] and Yuan [146]. Wang et al. [147] further studied the relation of two different substituent species in one cluster model and found that the strength of Brønsted acidity increases in the order of B-MCM-22 < [Al,B]-MCM-22 < Fe-MCM-22 < [Al,Fe]-MCM-22 < Al-MCM-22. Jittima [148] studied isomorphously substituted zeolite by the DFT method and observed that the trend of relative Brønsted acidity is [B]-LTL < [Ga]-LTL < [Al]-LTL, which is in agreement with the experimental sequence of isomorphously substituted ZSM-5 zeolite. Yang et al. [149] studied the Lewis and Brønsted acidities of various M^{4+}-doped zeolites (M = Ge, Ti, Pb, Sn, and Zr) and their interactions with probe molecules by the DFT method. According to the adsorption energies they obtained, the Lewis acidities of the zeolites increase in the order of silicalite-1 < Ge < Ti < Pb < Sn < Zr. As well, based on energy differences, interaction energies, and geometrical parameters, the Brønsted acidities increase in the order silicalite-1 < Ti < Ge < Zr = B < Pb < Sn < Al.

As mentioned above, the following conclusions may be drawn: [1] Low Al concentration can create strong acidic sites in zeolite; [2] The different Al sites affect the acid strength and activity of the zeolite; [3] The presence of isomorphously substituted metal ions exerts an important influence on the acidic properties of the zeolite. Our previous investigations were based on DFT methods, and the most important finding is that the simulation results are in good agreement with the experiments. Such findings confirm the reliability and importance of theoretical calculations, which can provide useful interpretations of experiment phenomena at the atom level. Theoretical calculations may also guide experimental procedures, thereby confirming their importance in investigations of various materials including catalysts.

5 Applications in Reactions of Refining Processes

5.1 Desulfurization

The catalytic removal of impurities, such as sulfur and nitrogen from oil products, via hydrotreatment is used to improve the quality of the final products. In particular, organosulfur compounds are present in different fractions in crude oil distillations. The low-boiling crude oil fraction mainly contains aliphatic organosulfur compounds, such as mercaptans, sulfides, and disulfides, which are very reactive in conventional hydrotreatment processes and can easily be completely removed. In high-boiling crude oil fractions, including heavy straight-run naphtha, straight-run diesel, and light FCC naphtha, thiophene and thiophene derivatives are the main components that are highly resistant to desulfurization [150].

5.1.1 Hydrodesulfurization (HDS)

HDS is one of most significant catalytic processes in the petroleum refining industry in removing sulfur from compounds containing thiophenic rings to generate clean fuels. HDS proceeds via two reaction pathways. The first one involves the direct removal of the sulfur atom from the molecule; this process is called direct desulfurization (DDS). The second pathway involves the hydrogenation of the aromatic ring and the subsequent removal of sulfur; this process is called hydrogenation (HYD). DDS and HYD usually occur simultaneously and the dominance of one process is dependent on the nature of sulfur compounds, the catalyst used, and other factors. Molybdenum sulfide (MoS_x) supported on γ-alumina is one of the main active components in hydrodesulfurization catalysts. Numerous studies have focused on the properties of the promoted MoS_x active phase, the support γ-alumina, as well as the interaction between them. The non-exhaustive list of experimental techniques that can be used to characterize the catalysts includes transition electron microscopy (TEM) [151, 152], X-ray photoelectron spectroscopy (XPS) [153, 154], IR spectroscopy [155, 156], and extended X-ray absorption fine structure (EXAFS) [157, 158]. EXAFS data provide valuable information about the local distances and coordination at the interface between the active phase and the support. However, establishing an atomistic representation of the interaction between the active phase and the support is still considerably difficult. Therefore the DFT and its implementation in efficient software programs, such as $Dmol^3$ [159], VASP [160] and CASTEP [161], can help alleviate this difficulty because the DFT models of HDS catalysts can be used to create an atomistic representation of the said interaction, and this strategy has shown considerable success.

During hydrogenation, the nonpromoted MoS_2 exhibits two active edges, including the ($10\bar{1}0$) molybdenum edge and the ($\bar{1}010$) sulfur edge, in comparison with the inactive (001) basal plane [162–165]. In contrast to periodic slab calculations, in which the average of the surface energies of the Mo-edge and S-edge [162] can be obtained, triangular-shaped clusters allow for an accurate determination of the surface energy of each individual edge. The Mo-edge is energetically the most stable surface under realistic HDS or working conditions [164]. In previous studies, the creation of S-vacancies in such clusters at the corners of the crystal is much easier than that at the edges, thereby indicating that the crystal corner affects the catalytic properties of nonpromoted MoS_2 particles. Paul and Payen [165] studied the formation of sulfur vacancies, and the results of their detailed analysis showed that the creation of one sulfur vacancy is kinetically more favorable on the Mo-edge (with 50% sulfur coverage) than on the S-edge (with the same coverage) [166].

Several electronic properties of the coordinately unsaturated Mo-edge sites (CUS) can be explained by analyzing the projected density of states close to the Fermi energy based on DFT simulations. The occupied d-states at the Fermi energy are related to the metallic character of Mo, whereas the unoccupied d-state above the Fermi level could be correlated with the adsorption energy [162, 167–169]. The adsorption mode of sulfur compounds [170], as a function of S-coverage, can affect the HDS reaction pathways based on DFT simulations. Compared with η^1 (S) (where S bonds to the Mo site), η^5 (the π-bonding mode) is more complex because of π-backdonation. The Mo atom with vacant d orbitals accepts π-electrons from the conjugated C=C double bond, thereby resulting in formation of strong π-bonding modes. This process is described as π-donation. Meanwhile, the vacant π* orbital of C=C accepts electrons from the occupied d orbitals of Mo; this process is described as π-backdonation. This donation/backdonation can considerably activate sulfur compounds and further induce S–C bond cleavage. These studies also imply that the formation of CUS is very important for HDS in a nonpromoted system even at low edge concentrations.

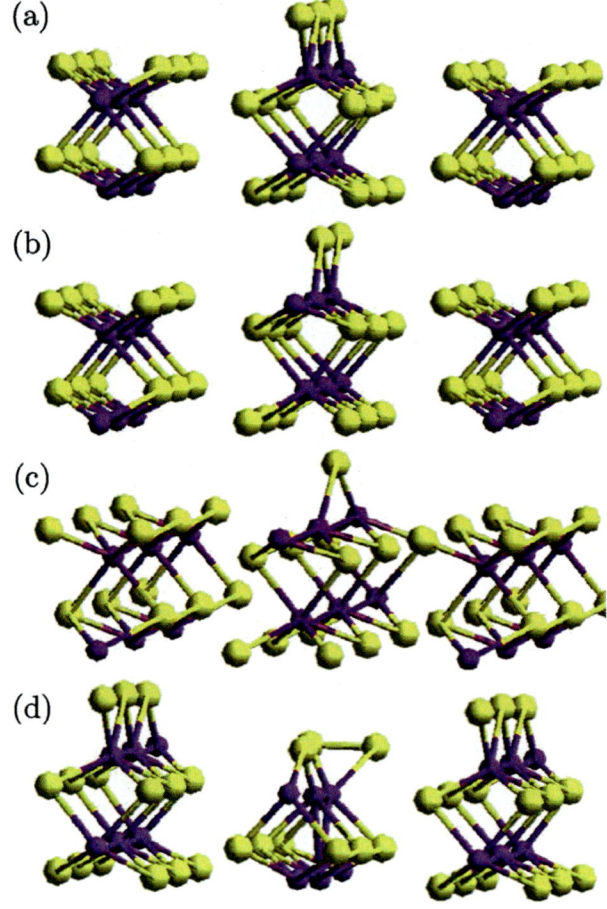

Fig. 10 The MoS$_2$ [100] surface. (**a**) Most stable surface. (**b**) Site 1. (**c**) Site 2. (**d**) Site 3 (sites 1, 2, and 3 reflect the removal of one, two, and three sulfur atoms, respectively, to create different vacancies)

Cristol et al. [170] investigated the adsorption mechanisms of some sulfur compounds over MoS$_2$ models with different vacancies (Fig. 10); their results indicated that η^1 (S) and benzene adsorption are competitive in dibenzothiophene (DBT) with a larger vacancy and that dimethyldibenzothiophene (DMDBT) is mainly adsorbed through its ring. This mechanism is very different from the adsorption behaviors of benzothiophene (BT) and 4-methylbenzothiophene (MBT). This group attributed the difference observed to the aromaticity of DBT and DMDBT and the fact that steric hindrance prevents adsorption through the thiophene ring. A mechanistic study of BT hydrodesulfurization on Mo$_3$S$_9$ model clusters indicated that the hydrogenation and hydrogenolysis of BT with the intermediate DHBT are more favored than the formation of the intermediate styrene [171]. The increased HDS reactivity of thiophene is due to the direct formation of butadiene with the intermediate 2,5-DHT [172]. Hydrogenolysis of C$_2$H$_5$SH to form C$_2$H$_6$ is much easier than its elimination to form C$_4$H$_4$ because the energy barrier associated with the former is lower and C$_2$H$_5$SH shows higher activity than

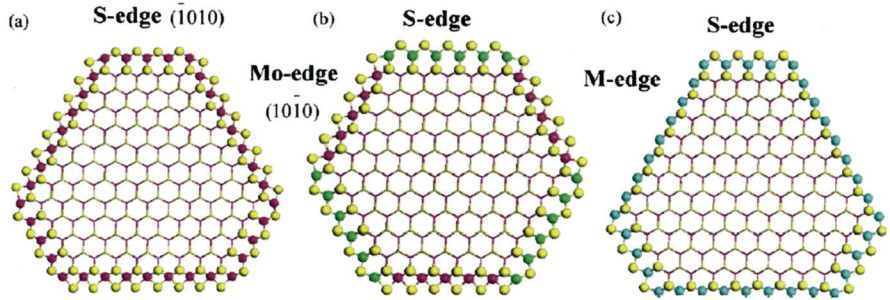

Fig. 11 Calculated DFT equilibrium morphologies at HDS conditions of the (**a**) nonpromoted MoS_2 active phase, (**b**) CoMoS with Co at the S-edge only, and (**c**) NiMoS with Ni at the S-edge and Mo-edge [163–165]. Only one Mo-edge is substituted by Ni. *Yellow balls*, sulfur; *magenta balls*, molybdenum; *green balls*, cobalt; *blue balls*, nickel (to interpret the color references in this figure legend, refer to the web version of this article)

CH_3SH [173, 174]. However, the S–Mo bond energy is generally accepted as high at the edge, thereby limiting the adsorption of reactants and H_2 dissociation and further weakening the catalytic activity of the nonpromoted system.

Some transition metals added to the MoS_2 can increase the reactivity of the resultant catalysts. These transition metals are considered promoters rather than catalysts in their own right because of their small added fraction relative to that of Mo. Co and Ni are considered strong promoters of HDS activity [175, 176]. The volcano-type relationship between the calculated adsorption energy of thiophene and the experimental HDS activity indicates that Ni and Co are strong promoters, whereas other metals (V, Cr, Fe, Cu, and Zn) show moderate or weak promoting effects [177] because only the edges, and not the basal plane sites, are active for unpromoted MoS_2 [178] and incorporation of Co and Ni atoms can decrease the binding energy of sulfur on the edge planes. Thus, the equilibrium sulfur coverage on the edge surfaces of the promoted system is reduced to create more active sites [155, 168, 179–182].

The exact origin of the promotory function of Co and Ni and the atomic-scale location of Co and Ni have been gradually explored and characterized via experimental techniques [183–187]. EXAFS data provide relevant insights into the local environments of the promoter atoms (Co or Ni) in the structure; these data include Co(Ni)–Mo distances of 2.75 Å–2.90 Å and Co(Ni)–S distances of 2.10 Å–2.20 Å [158, 188, 189]. Co is more stable during substitution at the S-edge than that at the Mo-edge because the former has a smaller edge energy [164, 168, 190]. Figure 11 shows the morphologies of MoS_2 and the Co(Ni)-promoted MoS_2 system [163–165].

Co substitution results in a nearly hexagonal morphology compared with the triangular morphology of unpromoted MoS_2. Although Ni substitution is dependent on cluster size, in which larger Ni–Mo–S particles show a truncated triangular shape similar to Co–Mo–S nanoclusters, smaller Ni–Mo–S particles feature a dodecagonal shape, as shown in Fig. 12.

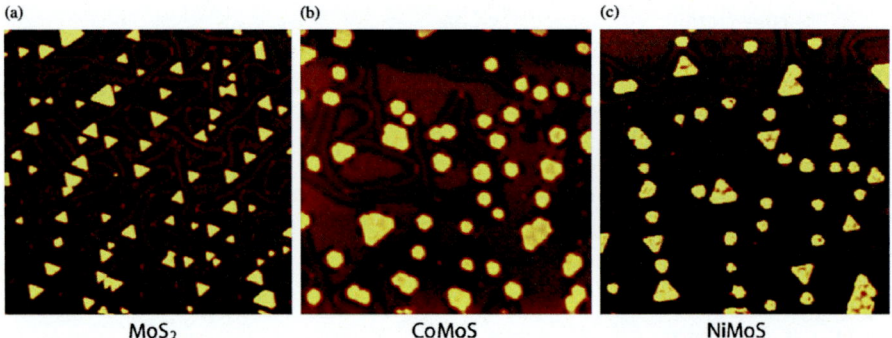

Fig. 12 (a) Morphology of unpromoted, Co-promoted, and Ni-promoted nanoclusters. (a) MoS$_2$ on Au [111] (700 × 700 Å^2). (b) Co–Mo–S on Au [111] and 700 × 700 Å^2. (c) Ni–Mo–S on Au [111] 700 × 700 Å^2 [165]

Reports concerning the HDS reaction of organosulfur compounds on promoted MoS$_2$ are scarce. Todorova et al. found that promoted atoms (Co and Ni) on the catalyst surface weaken the bond of adsorbed CH$_3$S and lower the energy barrier for CH$_4$ formation [174]. We hope that more researchers will devote their time to study the HDS mechanism with DFT and provide detailed insights into the promoted catalyst system.

A number of interactions between the active phase and the support (γ-alumina and anatase TiO$_2$) should be examined but remain unclear [191–193]. If interactions such as those of Mo–O–Al(Ti) or Mo–S–Al(Ti) are present in the system and connect the active phase with the support, it sounds unreasonable that only MoS$_2$ or Co(Ni)/MoS$_2$ is regarded as active phase but ignoring effect of support. Therefore, DFT simulations should be done carefully. More investigations about the nature of the active phase and support are required.

5.1.2 RADS

Reactive adsorption desulfurization (RADS) obtained great concerns because it has the ability to remove the last ppm sulfur in the fuel production [194–198]. During RADS, the sulfur atom is removed from a sulfur-containing molecule, such as thiophene, and benzothiophene and then fixed onto an adsorbent. The S-free hydrocarbon is then returned to the final oil product or further hydrogenated to form saturated hydrocarbons in a hydrogen atmosphere. Thus, the negative impact of RADS on the yield of liquid product and octane number is low because only the sulfur atom is removed. RADS has been implemented in the S-Zorb process of ConnocoPhillips Petroleum Co. to remove sulfur from gasoline and diesel [199–202]. The main compositions of the S-Zorb sorbent contain alumina, silica, and nickel and zinc oxides. Different catalysts have been tested, and results consistently demonstrate that the Ni/ZnO system presents excellent catalytic performance. In

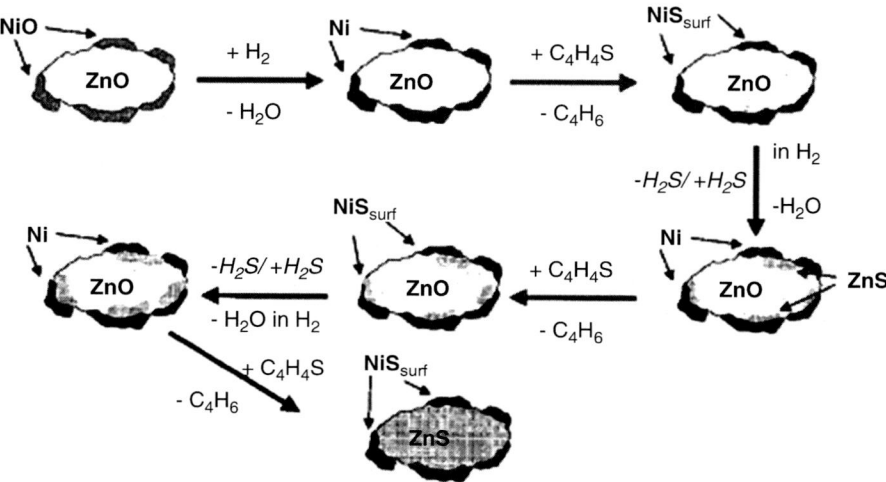

Fig. 13 Mechanism of reactive adsorption desulfurization

such a system, ZnO provides a high sulfur capacity and Ni promotes the decomposition of organic sulfur compounds [203, 204]. Babich and Moulijn [150] tentatively proposed a reaction scheme described in Fig. 13 wherein thiophene is first decomposed on the Ni/ZnO surface followed by reduction of nickel sulfide in the presence of H_2 to regenerate active nickel. H_2S is then absorbed by ZnO, which is transformed into ZnS. Many experiments [196, 205–209] have been conducted to investigate the mechanisms RADS but the detailed desulfurization process remains unclear because of various experimental limitations.

DFT simulations have been gradually applied to study the desulfurization mechanism at the micro level. Thiophene is often chosen as the model organic sulfur compound because it has a lower reactivity than alkyl thiophenes and benzothiophenes in the reactive adsorption process [210]. During the desulfurization reaction, the initial and key step is the adsorption of thiophene on the sorbent. The adsorption described here is different from the definition of adsorption desulfurization (ADS), which is performed to remove the entire sulfur-containing molecule based on physical interactions between the organosulfur compounds and the sorbent. Many π-complexation adsorbents with different selectivities and adsorption capacities [211–214], such as Cu(I)–Y zeolite and Ag–Y zeolite, are used in ADS, in which various species are further removed from liquid fuels after HDS based on their competitive adsorption abilities. Several sorbents have also recently been applied to ADS, including microporous coordination polymers [215], activated carbon [216], and MOFs [217]. Adsorption through RADS also involves chemical interactions between the organosulfur compounds and the active metals. The molecular and dissociation adsorption of thiophene can occur at various interaction strengths with metals [218].

Generally, several main adsorption mechanisms of thiophene with metals including the η^1 (the S bond to the metal site), η^2 (thiophene π-binding to a metal

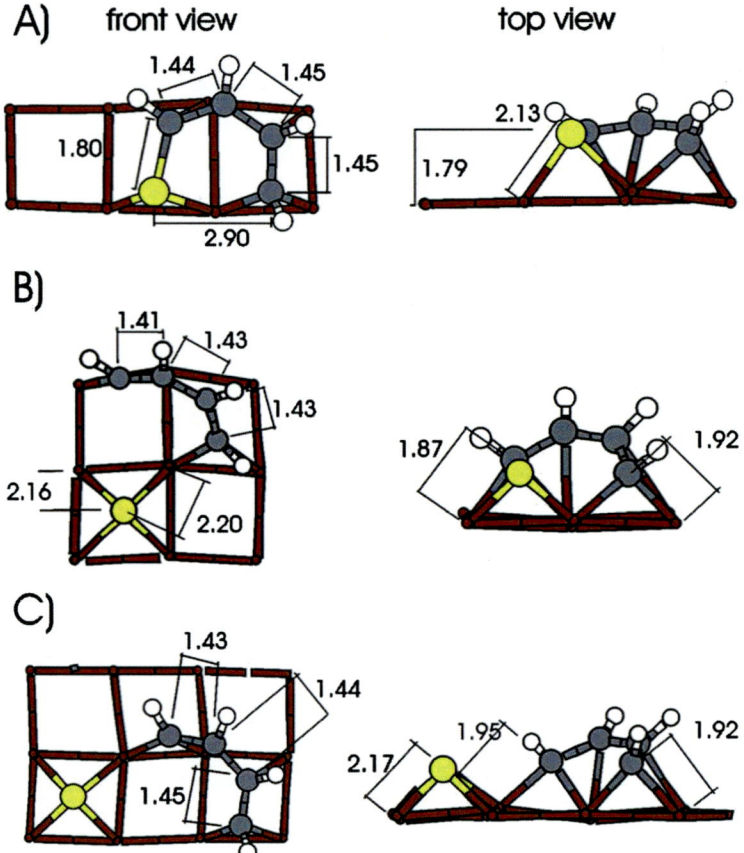

Fig. 14 Thiophene adsorbed on a Ni [100] surface: adsorption geometry (**a**) at the initial bridge position with one broken S–C bond, (**b**) after the second S–C bond scission and the diffusion of the sulfur atom into the neighboring hollow site, and (**c**) after the second S–C scission and sulfur diffusion into an alternative hollow site

site through one C=C double bond), and η^5 bonding modes (thiophene π-binding to a metal site through all five atoms in the ring) have been considered. Configurations with the η^1 bonding mode are less energetically favorable compared with those with the two other bonding modes [219–222]. Moreover, by comparing the adsorption energies and analyzing the electronic properties of the metals, the stable adsorption geometry can be determined. After thiophene adsorption, charge distribution again occurs between thiophene and the metals. For the η^1 bonding mode, electron transfer from thiophene to active metals results in a positive charge in the thiophene ring. In general, a large value of charge transfer indicates a powerful adsorbent–adsorbate interplay. The η^2 and η^5 adsorption modes are highly stable because of the π-donation/backdonation process (see previous section). When the charge density maps of the adsorption systems are analyzed, the charge transfer between

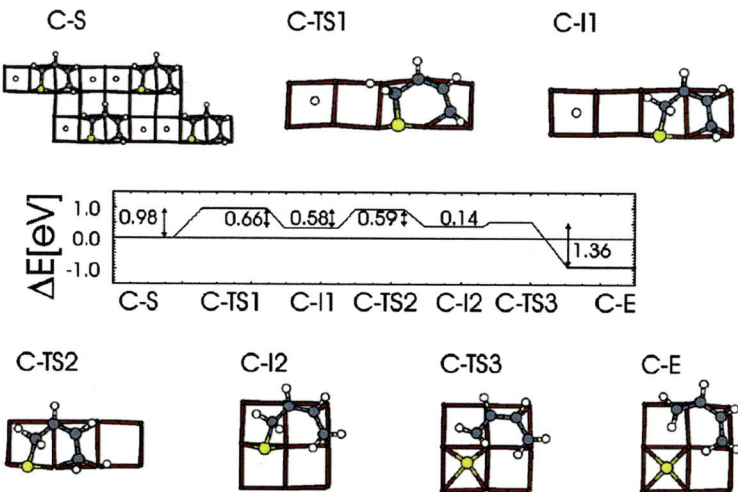

Fig. 15 Energetic profile of the reaction pathway for the hydrodesulfurization of thiophene/Ni [100]: starting from coadsorbed thiophene and hydrogen (C–S), the first transition state (C–TS1) for the initial hydrogenation of thiophene at the second position, a reaction intermediate (C–I1), the following TS for the hydrogenation of the fifth position (C–TS2), SC_4H_6/Ni [100] (C–I2), the final S–C bond scission (C–TS3), and coadsorbed S and C_4H_6 (C–E)

the adsorbate and the adsorbent can be understood and the transferred electrons of the adsorption system can be quantified by analyzing the Mulliken and Bader charges [52, 160, 219, 223, 224].

Mittendorfer and Hafner [225, 226] studied the adsorption of thiophene on Ni [100] via ab initio local-functional-density calculations and further described the thiophene desulfurization process, which ends with coadsorption of S and C_4H_4. In their investigations, thiophene adsorption brought about the cleavage of one of the S–C bonds, and the sulfur atom was located at the bridge position (Fig. 14a). This configuration is due to the occupation of the antibonding thiophene π^*-orbitals caused by adsorption, which decreases the strength of the C–S bonds. Elongation of the remaining S–C bond results in the removal of a sulfur atom into the favored hollow site (Fig. 14b). An alternative reaction path involves the other neighboring hollow sites as the final position for the sulfur atom (Fig. 14c). Based on products B and C, two reaction pathways, A and B, were designed and the active barrier E_{reb} was calculated. Results showed that pathway B ($E_{reb} = 0.71$ eV) is slightly favored over pathway A ($E_{reb} = 0.84$ eV) because of steric reasons. Mittendorfer and Hafner also found that hydrogenolysis of the adsorbed thiophene molecule can lower the barrier of C–S scission by weakening the remaining sulfur–carbon bond and decreasing the energetic cost for thiophene molecule deformation (Fig. 15). However, the E_{reb} of the initial step in thiophene hydrogenation is 0.98 eV, which is higher than that observed for thiophene desulfurization (0.84 and 0.71 eV). As such, the former is not a favorable pathway. The adsorption and dissociation of thiophene on Ni [110] show that the most stable molecular adsorption structure corresponds to the highest energetic barrier for the complete decomposition of thiophene into S and C_4H_4 [160].

However, theoretical studies on the entire RADS process involving thiophene are fairly limited. Recently, the RADS of thiophene was tentatively described using cubane-like bimetallic oxide cluster models [227]. After the adsorption of thiophene on Ni, the generated nickel sulfides were either reduced in the presence of H_2 to form H_2S or directly transferred the sulfur atom from the Ni site to Zn site. The active barrier with H_2 is 0.859 eV but 0.634 eV for direct sulfur transfer. Thus, the latter is favored. However, hydrogen is required in RADS experiments and has prompted us to investigate the function of H_2 in the reactive adsorption of thiophene. Moreover, several crucial and intrinsic questions remain unanswered. Is there some interaction between Ni and ZnO? If yes, how does this interaction assist the reaction of thiophene with the adsorbents? How is the thiophene sulfur transferred and finally stored in the form of ZnS? What is the effect of the various dispersion distributions and configurations of Ni over ZnO on thiophene desulfurization? The lack of answers to these questions leads to the absence of a fundamental grasp of the intrinsic reaction mechanism, thereby directly limiting the rational design of novel and highly efficient catalysts and the further development of the RADS process.

5.2 Hydrodenitrogenation (HDN)

Nitrogen-containing compounds can cause catalyst poisoning, unstable products, and environmental pollution because NOx is emitted from exhaust gases [228]. As legislations and regulations of environmental protection become stricter, catalysts must show improved activity and selectivity for the HDN process.

In heavy crude oils, most nitrogen components present in the form of heterocyclic organonitrogen compounds, which contain five-numbered (non-basic) pyrrolic rings and six-numbered (basic) pyridinic rings. Non-basic nitrogen compounds make up a significant fraction of the total nitrogen content in heavy oils. However, these compounds are unstable and susceptible to oxidation and polymerization and inhibit hydrodesulfurization reactions [229]. Non-basic nitrogen compounds are also more difficult to remove compared with basic nitrogen compounds when conventional NiMoS catalysts are used. Will Kanda et al. [230] found that in a narrow boiling fraction of coker gas oil, non-basic carbazoles and tetrahydrocarbazoles are the primary species responsible for the higher inhibition and deactivation observed in the lightest fractions during the HDN of quinoline. The different reaction characteristics are attributed to the weaker adsorption of non-basic nitrogen compounds than basic nitrogen compounds [231, 232].

The main factors affecting the reactivity of nitrogen compounds include the distinct electronic structures and properties of non-basic and basic nitrogen compounds. Sun et al. [233] investigated the electronic structures and properties of representative organonitrogen compounds in crude oil via DFT, as shown in

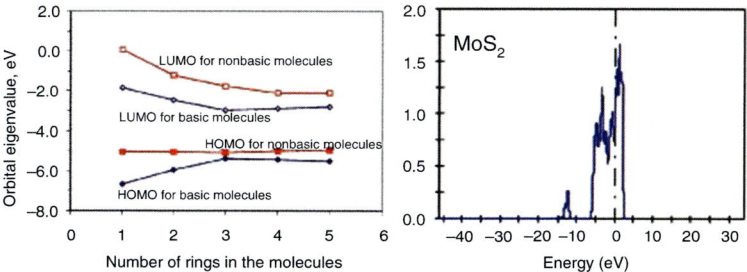

Fig. 16 Correlating the electronic properties and HDN reactivities of organonitrogen compounds: an ab initio DFT study [233]

Fig. 16; they found that during hydrogenation of aromatic molecules, the reactivities observed are related to the energy levels of occupied or unoccupied π-orbitals. Moreover, the occupied orbitals can donate electrons to unpromoted molybdenum catalysts because the active sites of the catalyst are strong electron-accepting sites. In Ni-promoted catalysts, unoccupied orbitals are crucial because the active sites of the catalysts are electron-donating sites and not electron-accepting sites. Thus, non-basic nitrogen compounds have higher energy π-orbitals than basic nitrogen compounds. Basic compounds may be more reactive than non-basic nitrogen compounds on a catalyst surface with a low density of unoccupied states, whereas the opposite may be true on a catalyst surface with a high density of occupied states.

Sun et al. [233] further investigated the adsorption and first-step hydrogenation of pyridine and pyrrole on the Ni-promoted (10.10) edge of MoS_2 via DFT calculations. They found that, for the hydrogenation of pyridine via a Langmuir–Hinshelwood mechanism, the hydrogen from the adsorbed H_2S is involved in the lowest-activation-energy reaction pathway and the hydrogen from the –SH groups is involved in the lowest-activation-energy reaction pathway for pyrrole at the edge of the MoS_2 (0001) basal plane. Eley–Rideal reaction pathways, which involve gas-phase pyridine or pyrrole and surface hydrogen species, require very low activation energy. Thus, the rate-determining step under these reaction conditions is the dissociation of hydrogen on the catalyst surface.

The adsorption of high-molecular-weight organonitrogen molecules such as quinoline, indole, acridine, and carbazole, which present a larger reaction resistance in heavy oil, on the active nickel-promoted NiMoS edge surface must be studied. Sun et al. [234] found that quinoline and acridine (basic) are preferably adsorbed on the Ni-edge surface through the lone pair electrons of the nitrogen atom. This process produces relatively high adsorption energies. However, indole and carbazole (non-basic) primarily interact with the NiMoS catalyst edge surface through the p-electrons of the carbon atoms. The high HDN activities of Mo(W) carbide- and nitride-based catalysts have been investigated. Among the carbides of early transition metals (TMCs), hexagonal Mo_2C bulk and supported catalysts are most active in the HDN process [235]. Generally, hydrogenation of the heteroatom ring occurs prior to C–N bond breaking. The HDN reaction proceeds in two parallel

Fig. 17 Schematic network of consecutive-parallel reactions used for the kinetic modeling of HDN of indole [235]

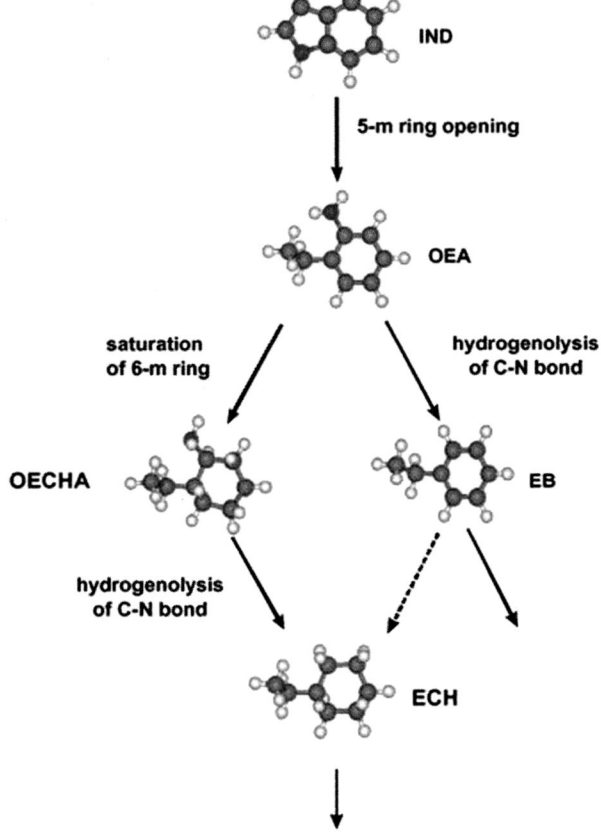

routes: direct (DDN) or indirect (HYD), where nitrogen is removed depending where hydrogen attacks, that is, either the heterocyclic or carbocyclic ring, respectively.

The scheme for isothermal kinetic studies on indole HDN over Mo_2C catalysts is shown in Fig. 17.

Witold Piskorz et al. [236] investigated the molecular reaction mechanism of indole HDN over Mo_2C via DFT and found out that the calculated difference in energy barriers (8 kcal mol^{-1}) between the DDN and HYD routes provides a molecular rationale for the observed higher production rates of aromatics (EB) over those of aliphatics (ECH).

Hiroyuki Tominaga et al. [228] investigated the HDN of carbazole on a γ-Mo_2N (1 1 0) slab via DFT calculations, as shown Fig. 18. The nitrogen atom of the carbazole is adsorbed and hydrogenated on the Mo atom of the γ-Mo_2N(1 1 0). On the bridge position of two Mo atoms, the hydrogen molecule is dissociatively adsorbed. The hydrogenation of tetrahydrocarbazole to octahydrocarbazole proceeds via hydrogen attack on the carbon atoms. This attack on the carbon atoms

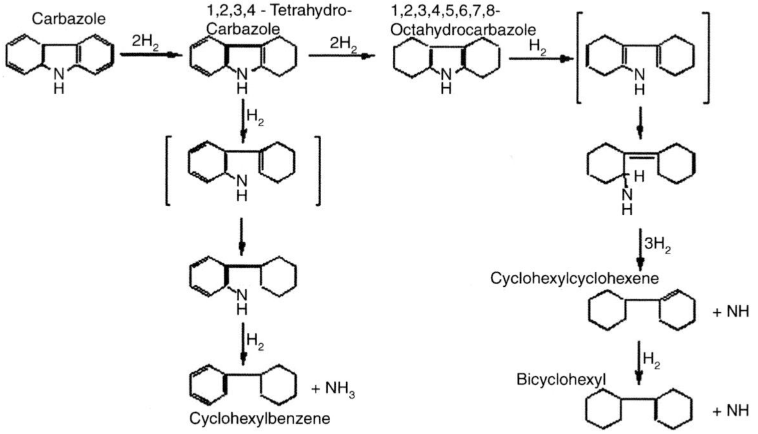

Fig. 18 Reaction mechanism for the hydrodenitrogenation of carbazole on molybdenum nitride via DFT

close to the nitrogen atom results in scission of the C–N bond of tetrahydrocarbazole and octahydrocarbazole because the activation energy is low. Thus, the conformation of the adsorbed compound changes when the height from the top Mo surface is reduced. The C–N bonds of the tetrahydrocarbazole are cleaved, and cyclohexylbenzene is finally formed via 1-cyclohexyl-2-iminobenzene. Octahydrocarbazole is not hydrogenated to perhydrocarbazole, but one C–N bond is cleaved to form bicyclohexyl through cyclohexylcyclohexene. One C–N bond of the hydrogenated carbazoles is cleaved on the Mo atom, but denitrogenation occurs at the bridge position of the two Mo atoms [228].

Understanding the correlations between the electronic structures and properties of organonitrogen molecules and their catalysts is an important step, and determining the reaction mechanism ensures the successful development of excellent catalysts with high activity and selectivity for use in HDN reactions.

5.3 Alkylation

Alkylation, an efficient method to produce gasoline blending components with a high octane number, has become a very important industrial process. Commercial alkylation processes are performed using sulfuric acid or hydrogen fluoride as catalysts. The desired products, namely, highly branched C8 alkanes, are mainly trimethylpentanes (TMPs), whereas the undesired products are DMH (dimethylhexane), which do not increase the octane number. Thus, the relevant reaction mechanisms should be determined to improve the production of TMP and enhance the commercial benefit of this process. Wang et al. [237, 238] investigated the reaction path of 2-butene and isobutane via DFT and ab initio method. They

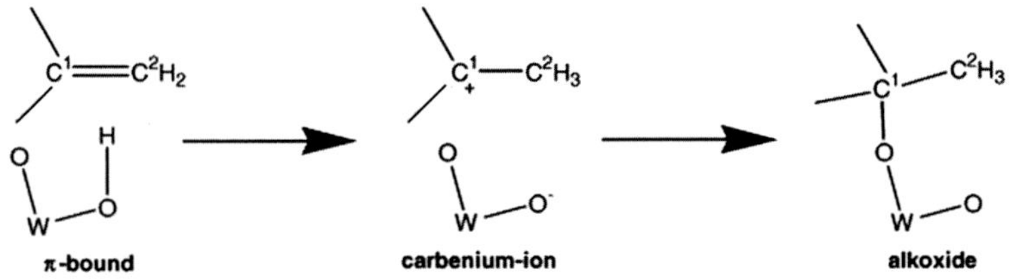

Fig. 19 Mechanism of alkene adsorption over phosphotungstic acid. The tungsten-oxide fragment represents a section of the Keggin unit

analyzed and compared the geometrical structure and energy change and found that 2-butene is rapidly protonated to form a sec-butyl carbonium ion, which then reacts with isobutane to form a *tert*-butyl carbonium ion. This carbonium ion is then deprotonated to form isobutene, which rapidly reacts with *tert*-butyl carbonium ion to form a trimethylpentane (TMP) carbonium ion. Finally, the TMP$^+$ carbonium ion is converted to TMP via the hydride transfer process. The authors also found that the energy barrier of each reaction is below zero, thereby suggesting that each reaction step is a rapid carbonium-ion reaction step. Homogeneous catalysts, such as HF and H_2SO_4, are highly corrosive and cause waste disposal and catalyst separation problems. Thus, HF and H_2SO_4 are not environment-friendly catalysts. In recent years, many new types of alkylation catalysts, such as superacid, solid acid, and ionic liquid (IL), have been discovered.

Studies on solid isobutane alkylation catalyst, zeolites [239], mesoporous materials [240, 241], and heteropolyacid [242] have been ongoing for several decades. Although many strong solid acids have been tested for use as an alkylation catalyst that is viable, low cost, and environment-friendly, no process can replace traditional ones involving HF and sulfuric acid-based technologies [243] because of catalyst deactivation. A more thorough understanding of the fundamental structural and electronic features, which determine the catalyst acidity, selectivity, and potential modes of deactivation, is required. Wang et al. [244] proposed that methanol can compete with olefins and thiophenic sulfurs for acidic active sites on the catalyst because of its strongest adsorption strength, as determined via the DFT method. Methanol has an advantageous effect on HY zeolite performance during the desulfurization process. Janik et al. [245] examined the energetics (or energies) of key reaction steps during the alkylation over phosphotungstic acid via ab initio DFT methods. The adsorption of alkenes to HPAs is initiated by the formation of a π-bound state, as illustrated in Fig. 19. The authors further revealed that the energy barrier for alkylation is intrinsically lower than that for hydride transfer, thereby providing a favorable path for the buildup of heavy hydrocarbons on the acid surface.

Compared with solid acid catalysts, IL catalysts have more tunable properties. The organic cation and inorganic anion in ILs can have very different physical and chemical properties [246–248]. Chloroaluminated ILs and protic ammonium-based ILs have been extensively investigated for alkylation [247, 249]. Liu et al. [250, 251] synthesized a series of ILs containing different anions or cations

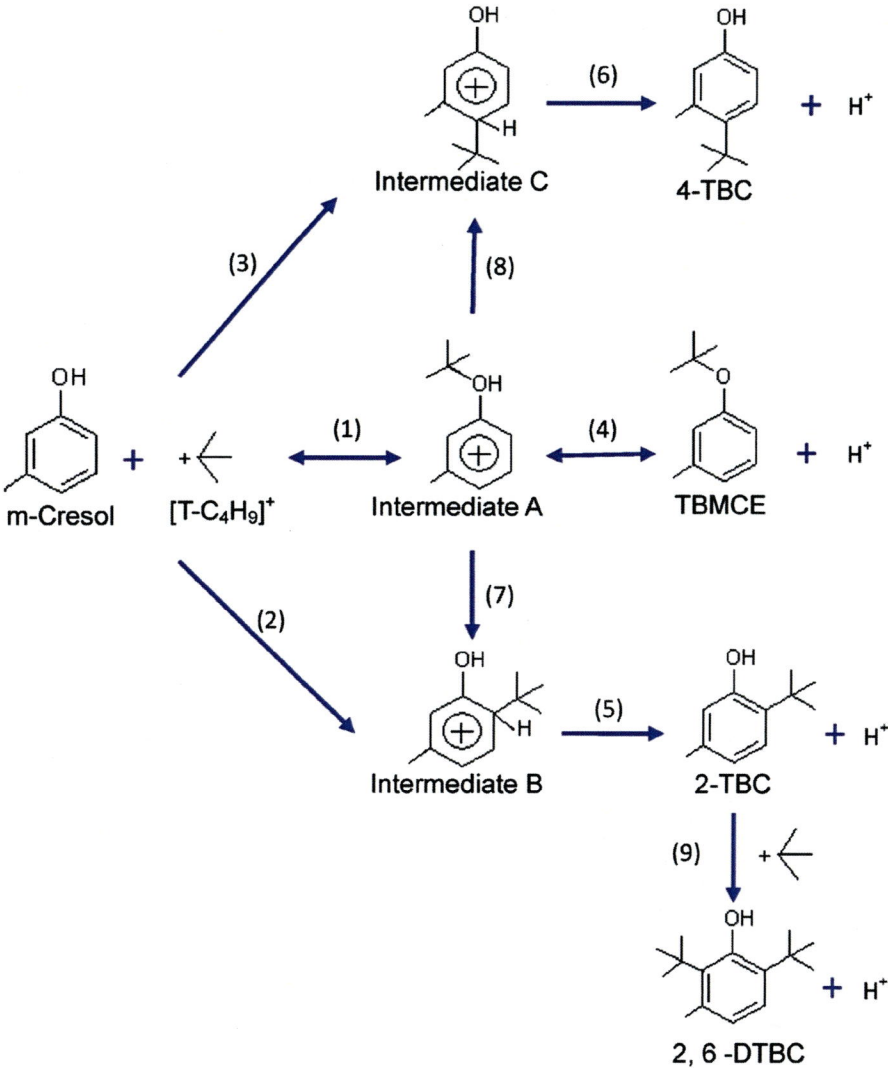

Fig. 20 Proposed reaction network for the alkylation of *m*-cresol

to investigate the alkylation of isobutane with butene. They found that $[Al_2Cl_7]^-$ is the origin of catalytic activity. After different ions of transition metals such as copper, iron, and zinc were dissolved in ILs, results showed that the catalytic properties of ILs with Cu^{2+} and Cu^+ are greatly improved along with C8 selectivity, which reached 75%, and the octane number of the alkylate, which reached 92.2 (RON). Room-temperature IL-catalyzed alkylation of isobutane with butene was performed in a pilot-plant scale, and the process was proven successful [252]. Later, they found that Et_3NHCl-$AlCl_3$ IL presents good catalytic performance for the isomerization of *n*-pentane [253, 254]. When the yield of isomerate and the selectivity of liquid isoalkanes increase, the cyclohexane amount, which is the optimal product distribution, also increases. Composite ILs (CILs) prepared from triethylamine hydrochloride, anhydrous aluminum(III) chloride, and cuprous chloride show efficient catalytic performance [255–258] and have gained increased attention from researchers. Liu et al. [258] found that CILs modified with copper (I) chloride are highly active and can be used as selective catalysts for the alkylation of 2-butene with isobutane. In addition to data obtained from experimental studies, the geometrical structures of Et_3NHCl, $AlCl_3$, and $[Et_3NH][AlCl_4]$ were successfully calculated via DFT at the B3LYP/6-31G*level[243]. Natural bond orbital (NBO) analysis shows that the atomic orbitals of Et_3NHCl and $AlCl_3$ have similar symmetries, thereby suggesting that the geometrical structure of $[Et_3NH][AlCl_4]$ is stable. Zhou et al. [259] investigated the reaction mechanism of the alkylation of *m*-cresol with *tert*-butanol catalyzed by a SO_3H-functionalized IL via quantum chemical calculations. The proposed reaction network for alkylation of *m*-cresol is shown in Fig. 20. The authors found that the selectivities of the products depend on the fundamental natures of the reactive sites, including the orbital overlap as well as Coulombic and steric effects on the interaction between the *tert*-butyl ion ($[t\text{-}C4H9]+$) and *m*-cresol.

5.4 Isomerization

In recent years, prompted by increasing environmental concerns worldwide, regulations on the sulfur, benzene, aromatic, and olefin contents of gasolines as well as their vapor pressure have become stricter, thereby leading to more stringent requirements for gasoline quality [260]. Desulfurization of gasoline results in a decrease in octane number; moreover, limitation of the vapor pressure and aromatic content leads to the drop of C4 and reformate. To maintain octane ratings, the amount of isomerate or the branched isomers of C5 and C6 must be increased in the gasoline pool [261].

The catalysts of most commercial isomerization units, such as platinum on aluminum chloride, are amorphous and resistant to feed purification processes; thus, these catalysts can cause corrosion problems and environmental hazards. In recent years, the use of zeolite has gained increased attention because the number of acid sites can be maximized and the characteristics of the channel structures inside the pores may be determined when shape-selective reactions occur in these structures [262, 263]. Compared with chlorinated alumina catalysts, zeolites are operated at a higher temperature, typically at 220–300°C, which is unfavorable for the thermodynamics of an ideal product [261, 263]. To understand and improve the activity and selectivity of the zeolite, numerous experiments must be performed to determine the feasibility of the use of zeolites for alkene isomerization [264–268]. However, several mechanism issues remain unresolved. Extensive studies on skeletal isomerization have been published, and the reaction is known to proceed

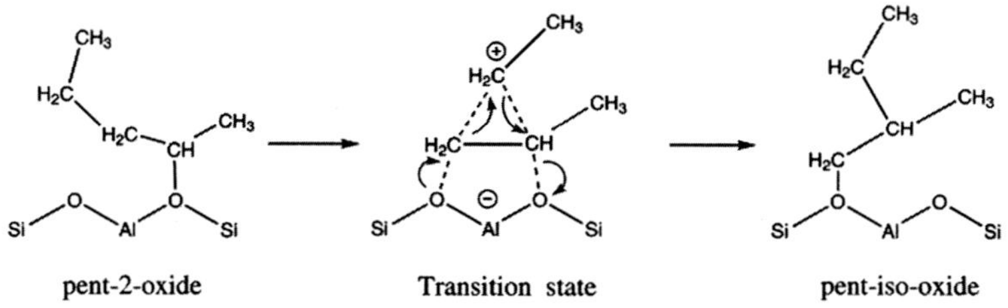

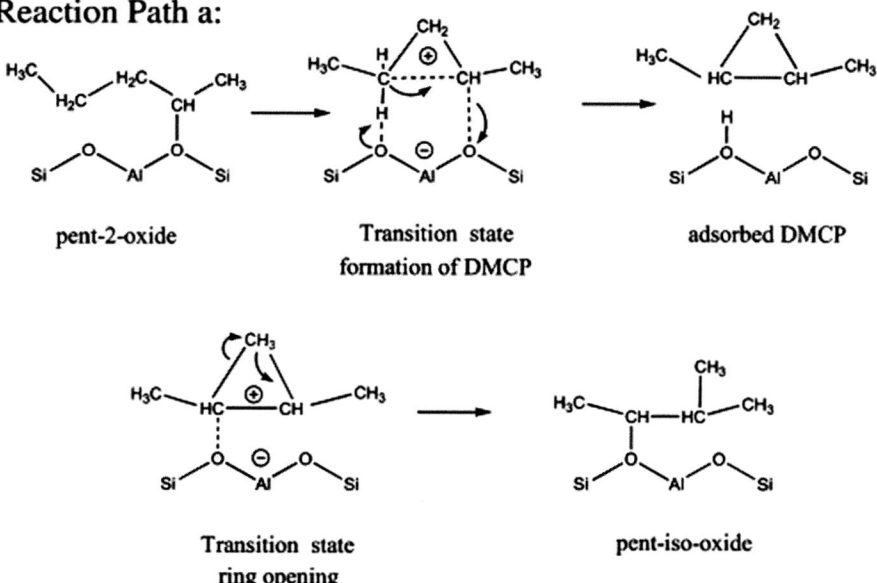

Fig. 21 Reaction mechanism of C5 isomerization catalyzed by acidic zeolites via an alkyl (ethyl) shift

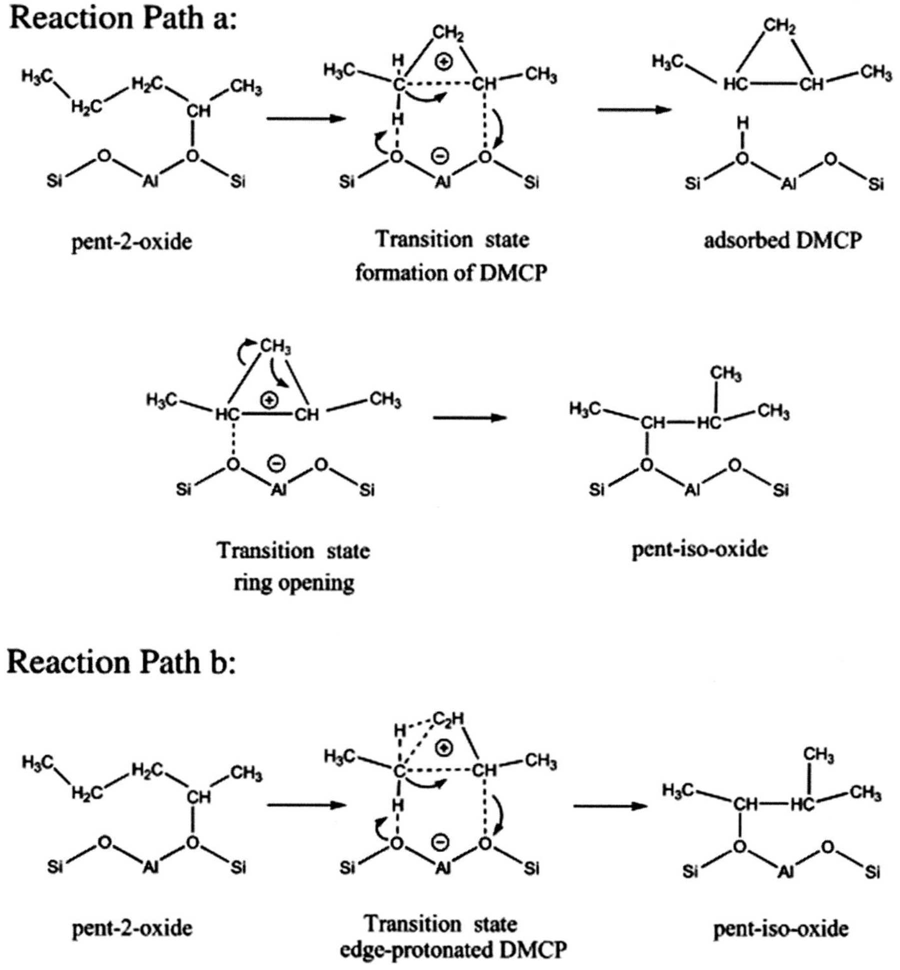

Fig. 22 Reaction mechanism of C5 isomerization catalyzed by acidic zeolites via a dimethylcyclopropane intermediate (path a) or via edge-protonated dimethylcyclopropane transition state (path b)

through either a mono- or a bimolecular reaction pathway [266, 269]. The monomolecular mechanism is the usual mechanism for *n*-pentane and larger hydrocarbons in zeolites and sulfated zirconia. For butene skeleton isomerization, the monomolecular mechanism is observed to occur commonly, and by-products are formed via dimerization, oligomerization, and cracking reactions [269].

Monomolecular isomerization might occur through two possible reaction pathways, namely, a one-step alkyl shift mechanism or a "protonated cyclopropane" intermediate reaction pathway [270], such as the reaction path of a pentene via an ethyl shift for C5 isomerization and the reaction scheme of the intermediate carbonium ion that are shown in Fig. 21.

Compared with C4 isomerization, C5 isomerization, especially that of pentene, is more complicated because the number of heterogeneous products produced from isomers is higher in the latter than in the former, and these by-products may react via double-bond isomerization. Thus, experimental studies cannot provide the detailed reaction steps, thereby hindering researchers from understanding the relevant micro mechanism. However, understanding the isomerization steps is imperative to develop more effective catalysts. Demuth et al. [270] investigated the skeletal isomerization of a 2-pentene molecule catalyzed by acidic ZSM-22 via ab initio DFT studies and found that the more likely pathway for skeletal isomerization inside the channels of ZSM-22 involves the rearrangement of the carbenium ion into a protonated dimethylcyclopropane ion. Such rearrangement implies the formation of relatively stable secondary carbonium ions as transient intermediates, shown as Fig. 22.

Yu-Hua Guo et al. [271] investigated the double-bond isomerization of 1-pentene over zeolites via DFT and found that double-bond isomerization may proceed via a stepwise or a concerted reaction pathway. The stepwise reaction involves two elementary steps: First, an alkoxy intermediate is formed by addition of a proton from the zeolites. This alkoxy intermediate then decomposes to form an adsorbed 2-pentene. The concerted reaction occurs via one-shift proton transfer, where the formation of highly stable alkoxide species is avoided. The double-bond isomerization of 1-pentene to *cis*-2-pentene occurs on the surface of the molecular sieves via a concerted proton transfer between the Brønsted acid sites of the zeolite and the pentene molecules [272].

Hong Li et al. [273] also investigated the isomerization of 1-hexene to 2-hexene on the surface of aluminosilicate molecular sieves via ab initio studies and found that 1-hexene can be directly transformed into 2-hexene by shifting the isomerization of the double bond because the energy barrier of the catalyst is very low over the aluminosilicate molecular sieve.

In fact, in order to obtain maximum economic benefits, the catalysts utilized in the isomerization industry are always bifunctional. "Bifunctional" catalysts are metallic catalysts, such as Pt or Pd, that are dispersed over an acidic zeolite catalyst and are used in isomerization. The well-dispersed metallic clusters allow equilibrium in the (de)hydrogenation step, and skeletal isomerization is the rate-determining step [238, 249, 263]. In recent years, anion-modified metal oxides, specifically sulfated zirconia (SZ), have gained increased attention. Sulfated zirconia has been the focus of studies since 1962 when it was first used as an isomerization catalyst. However, a major problem, the rapid deactivation of sulfated zirconia, is still unresolved. In order to overcome this problem, several transition metals, such as platinum, have been added as promoters to improve the activity and stability of the catalyst [238, 262]. The supercritical fluid of sulfated zirconia has also been applied to reduce the deactivation rate and achieve stable activity during solid acid-catalyzed isomerization. Takako et al. [274] found that Lewis acid sites on sulfated zirconia play an important role in the isomerization of n-butane. Sulfated zirconia is a solid super acid, and its strong acidity can cause rapid deactivation during the isomerization of alkanes. They also found that, at the supercritical condition, no obvious deactivation and the steady-state activity are maintained.

To obtain more information about the structure of the active species on the SZ surface for alkane isomerization, Tomonori Kanougi et al. [275] investigated the nature of the superacidity of the S-ZrO2 surface via periodic DFT study. The researchers found that H_2SO_4 dissociatively adsorbs on the tetragonal (1 0 1) ZrO_2 (t-ZrO_2) surface to produce H_2O and SO_3 molecules. In the case of Brønsted acid sites, the proton on H_2O and SO_3 coadsorbed on the t-ZrO_2 surface had a higher acidity than that of H_2O adsorbed on the t-ZrO_2 surface.

5.5 Hydrodeoxygenation (HDO)

From an experimental viewpoint, determining the HDO mechanism, which includes C=O hydrogenation and C–O bond cleavage, on the surface of MoS_2-based catalysts is difficult. Dupont et al. [276] evaluated the adsorption properties and determined the HDO pathways of a relevant model via DFT; this model includes O-containing molecules on the M-edge sites of the MoS_2 and NiMoS active phase, as shown in Fig. 23. They found that the bidentate adsorption mode of aldehyde on the Ni–Mo site promotes C=O hydrogenation into alcohol, which is further converted by C–O cleavage via nucleophilic substitution by sulfhydryl. Badawi et al. [277] investigated furan HDO catalyzed by MoS_2 via DFT calculations. During the reaction, the stability of the MoS_2 catalyst depends on the ratio of H_2S/H_2O, and the metallic edge is stable regardless of this ratio. To prevent the partial oxygenation of the sulfur located at the edge, the H_2S pressure must be maintained at a constant value. Once a vacancy has been created by removing a sulfur atom from the metallic edge, the adsorption of furan through its oxygen atom is possible. However, furan cannot be adsorbed on the stable surface.

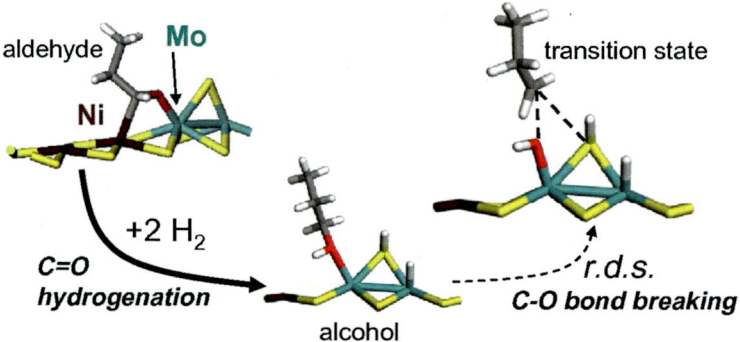

Fig. 23 Hydrodeoxygenation pathways catalyzed by MoS_2 and NiMoS active phases

6 Summary and Prospects

Based on the examples discussed, the molecular simulation methods were consistently confirmed to be a useful technique for understanding the adsorption and diffusion of adsorbate molecules and obtaining information on the design and prediction of the ultimate properties of materials. Given that experimental methods may become more expensive over time and that computational approaches present lower costs because of developments in computers, computational methods show great promise in the design and development of novel catalysts. Every microprocess at the atomic or molecule level can be studied on computers. Developing theoretical bases and methods of computation chemistry to improve the performance of HPC may lead to exciting possibilities in the field of chemistry, the chemical industry, and petroleum refining.

Acknowledgment This study was also sponsored by the National Natural Science Foundation of China (Grant Nos 21176253, 21476260, 21236009, 21336011, and U1162204).

References

1. Gubbins KE, Moore JD (2010) Ind Eng Chem Res 49:3026
2. Cohen AJ, Mori-Sanchez P, Yang WT (2012) Chem Rev 112:289
3. Waterman HI (1958) Correlation between physical constants and chemical structure: graphical statistical methods of identification of mineral and fatty oils, glass, silicones, and catalysts. Elsevier, Amsterdam
4. Van Nes K, Van Westen HA (1951) Aspects of the constitution of mineral oils. Elsevier, Amsterdam
5. Brown J, Ladner W (1960) Fuel 39:87
6. Liang WJ (2011) Petroleum chemistry. University of Petroleum Press, Dong Ying
7. Zhao L, Chen Y, Gao J, Chen Y (2009) J Mol Sci 311
8. Pan Y, Wang D, Gao J (2007) Acta Pet Sin (Pet Process Sect) 63

9. Zhao S, Kotlyar LS, Woods JR, Sparks BD, Hardacre K, Chung KH (2001) Fuel 80:1155
10. Gao J, Xu C, Kotlyar LS, Chung KH (2003) J Chem Ind Eng (China) 9
11. Duan A, Xu C, Gao J, Lin S, Chung KH (2005) J Mol Struct 734:89
12. Duan A (2003) J China Univ Pet (Sci Technol Ed) 91
13. Zhen K, Gao J, Xu C (2004) J China Univ Pet (Sci Technol Ed) 129
14. Wang D, Zhao Y, Pan Y, Liu R, Gao J (2006) J Fuel Chem Technol 690
15. Ren W, Chen H, Yang C, Shan H (2009) J Chem Ind Eng (China) 1883
16. Campbell DM, Bennett C, Hou Z, Klein MT (2009) Ind Eng Chem Res 48:1683
17. Campbell DM, Klein MT (1997) Appl Catal Gen 160:41
18. Wang C, Zhou H, Wang Z, Dai Z, Zhao Y (2012) Comput Appl Chem 1221
19. Ouyang F, Wang S, Jiang H, Weng X (2007) J Fuel Chem Technol 678
20. Ma F, Yuang Z, Weng X (2003) J Chem Ind Eng (China) 1539
21. Shen R, Cai J, Jiang H, Weng X (2005) J East China Univ Sci Technol 56
22. Zhang H, Zhao Y, Luo D (2011) J Chem Ind Eng (China) 705
23. Zhang H, Gu P, Nan Z, Shao Y (2012) Chem Prod Technol 34
24. Zhang LZ, Hou Z, Horton SR, Klein MT, Shi Q, Zhao SQ, Xu CM (2014) Energy Fuels 28:1736
25. Zhang L, Xu Z, Shi Q, Sun X, Zhang N, Zhang Y, Chung KH, Xu C, Zhao S (2012) Energy Fuels 26:5795
26. Zhang L, Zhang Y, Zhao S, Xu C, Chung KH, Shi Q (2013) Sci China Chem 56:874
27. Zhao X, Liu Y, Xu C, Yan Y, Zhang Y, Zhang Q, Zhao S, Chung K, Gray MR, Shi Q (2013) Energy Fuels 27:2874
28. Zhao X, Shi Q, Gray MR, Xu C (2014) Sci Rep 4:5373
29. Zhang Y, Zhang L, Xu Z, Zhang N, Chung KH, Zhao S, Xu C, Shi Q (2014) Energy Fuels 28:7448
30. Pacheco-Sánchez J, Zaragoza I, Martínez-Magadán J (2003) Energy Fuels 17:1346
31. Murgich J, Rodríguez J, Aray Y (1996) Energy Fuels 10:68
32. Alvarez-Ramirez F, Ramirez-Jaramillo E, Ruiz-Morales Y (2006) Energy Fuels 20:195
33. Ortega-Rodríguez A, Cruz S, Gil-Villegas A, Guevara-Rodríguez F, Lira-Galeana C (2003) Energy Fuels 17:1100
34. Zhang S-F, Sun LL, Xu J-B, Wu H, Wen H (2010) Energy Fuels 24:4312
35. Fuchs AH, Cheetham AK (2001) J Phys Chem B 105:7375
36. Smit B, Maesen TL (2008) Chem Rev 108:4125
37. Suraweera NS, Albert AA, Humble JR, Barnes CE, Keffer DJ (2014) Int J Hydrog Energy 39:9241
38. Smit B (1995) J Phys Chem 99:5597
39. Granato MA, Vlugt TJ, Rodrigues AE (2007) Ind Eng Chem Res 46:321
40. Demontis P, Suffritti GB, Fois ES, Quartieri S (1992) J Phys Chem 96:1482
41. Dubbeldam D, Calero S, Vlugt T, Krishna R, Maesen TL, Beerdsen E, Smit B (2004) Phys Rev Lett 93:088302
42. Dubbeldam D, Calero S, Vlugt T, Krishna R, Maesen TL, Smit B (2004) J Phys Chem B 108:12301
43. Beerdsen E, Dubbeldam D, Smit B, Vlugt TJ, Calero S (2003) J Phys Chem B 107:12088
44. Wender A, Barreau A, Lefebvre C, Di Lella A, Boutin A, Ungerer P, Fuchs A (2007) Adsorption 13:439
45. Calero S, Dubbeldam D, Krishna R, Smit B, Vlugt TJ, Denayer JF, Martens JA, Maesen TL (2004) J Am Chem Soc 126:11377
46. Garcia-Perez E, Dubbeldam D, Maesen TL, Calero S (2006) J Phys Chem B 110:23968
47. Granato MA, Vlugt TJ, Rodrigues AE (2007) Ind Eng Chem Res 46:7239
48. Takahashi A, Yang FH, Yang RT (2002) Ind Eng Chem Res 41:2487
49. Ban S, Van Laak A, De Jongh PE, Van der Eerden JP, Vlugt TJ (2007) J Phys Chem C 111:17241
50. Krishna R, Van Baten J (2007) Chem Eng J 133:121

51. Dzhigit OM, Kiselev AV, Rachmanova TA, Zhdanov SP (1979) J Chem Soc Faraday Trans 1 75:2662
52. Zheng X, Zhang Y, Huang S, Liu H, Wang P, Tian H (2012) Comput Theor Chem 979:64
53. Krokidas P, Skouras E, Nikolakis V, Burganos V (2010) J Phys Chem C 114:22441
54. Zang J, Nair S, Sholl DS (2011) J Chem Phys 134:184103
55. Santander JE, Tsapatsis M, Auerbach SM (2013) Langmuir 29:4866
56. Jänchen J, Stach H, Uytterhoeven L, Mortier W (1996) J Phys Chem 100:12489
57. Woods GB, Panagiotopoulos AZ, Rowlinson JS (1988) Mol Phys 63:49
58. Dunne J, Mariwala R, Rao M, Sircar S, Gorte R, Myers A (1996) Langmuir 12:5888
59. Vlugt TJH, García-Pérez E, Dubbeldam D, Ban S, Calero S (2008) Computing the heat of adsorption using molecular simulations: the effect of strong coulombic interactions. J Chem Theory Comput 4(7):1107–1118
60. Kondo M, Yoshitomi T, Matsuzaka H, Kitagawa S, Seki K (1997) Angew Chem Int Ed Engl 36:1725
61. Sun MS, Shah D, Xu HH, Talu O (1998) J Phys Chem B 102:1466
62. Choudhary VR, Mayadevi S (1996) Zeolites 17:501
63. Silva JA, Rodrigues AE (1997) AIChE J 43:2524
64. Barrer R, Bultitude F, Sutherland J (1957) Trans Faraday Soc 53:1111
65. Sheludko A (1967) Adv Colloid Interf Sci 1:391
66. Thamm H (1987) J Phys Chem 91:8
67. Barrer RM (1984) In: Olson D, Bisio A (eds) Proceedings of the 6th international Zeolite conference. Butterworth, Guildford, pp 870–886
68. Zeng Y, Ju S, Xing W, Chen C (2007) Ind Eng Chem Res 46:242
69. Thamm H (1987) Zeolites 7:341
70. Goj A, Sholl DS, Akten ED, Kohen D (2002) J Phys Chem B 106:8367
71. Mentzen BF (2005) C R Chim 8:353
72. Lachet V, Boutin A, Tavitian B, Fuchs AH (1999) Langmuir 15:8678
73. Elliott JR, Lira CT (1999) Introductory chemical engineering thermodynamics. Prentice Hall PTR, Upper Saddle River
74. Mellot CF, Cheetham AK, Harms S, Savitz S, Gorte RJ, Myers AL (1998) J Am Chem Soc 120:5788
75. Fleys M, Thompson RW (2005) J Chem Theory Comput 1:453
76. Deroche I, Gaberova L, Maurin G, Castro M, Wright P, Llewellyn P (2008) J Phys Chem C 112:5048
77. Barthomeuf D (1984) J Phys Chem 88:42
78. Sauer J, Ugliengo P, Garrone E, Saunders V (1994) Chem Rev 94:2095
79. Huang J, Long W, Agrawal PK, Jones CW (2009) J Phys Chem C 113:16702
80. Zhang J, Burke N, Yang Y (2012) J Phys Chem C 116:9666
81. Song L, Sun Z-L, Rees LV (2002) Microporous Mesoporous Mater 55:31
82. Zhu J, Mosey N, Woo T, Huang Y (2007) J Phys Chem C 111:13427
83. Fitch A, Jobic H, Renouprez A (1986) J Phys Chem 90:1311
84. Brémard C, Ginestet G, Le Maire M (1996) J Am Chem Soc 118:12724
85. Rungsirisakun R, Nanok T, Probst M, Limtrakul J (2006) J Mol Graph Model 24:373
86. Abrioux C, Coasne B, Maurin G, Henn F, Jeffroy M, Boutin A (2009) J Phys Chem C 113:10696
87. Uytterhoeven L, Dompas D, Mortier WJ (1992) J Chem Soc Faraday Trans 88:2753
88. Demontis P, Yashonath S, Klein ML (1989) J Phys Chem 93:5016
89. De Mallmann A, Barthomeuf D (1988) Zeolites 8:292
90. Auerbach SM, Henson NJ, Cheetham AK, Metiu HI (1995) J Phys Chem 99:10600
91. Jousse F, Auerbach SM, Vercauteren DP (2000) J Phys Chem B 104:2360
92. Jirapongphan SS, Warzywoda J, Budil DE, Sacco A (2006) Microporous Mesoporous Mater 94:358
93. Zheng H, Zhao L, Yang Q, Gao J, Shen B, Xu C (2014) Ind Eng Chem Res 53:13610

94. Cottier V, Bellat J-P, Simonot-Grange M-H, Méthivier A (1997) J Phys Chem B 101:4798
95. Jirapongphan SS, Warzywoda J, Budil DE, Sacco A (2007) Microporous Mesoporous Mater 103:280
96. Chen Z, Zhao S, Xu Z, Gao J, Xu C (2011) Energy Fuels 25:2109
97. Chen Z, Gao J, Zhao S, Xu Z, Xu C (2013) AIChE J 59:1369
98. Weisz P (1973) Chemtech 498
99. Sapre AV, Katzer JR (1995) Ind Eng Chem Res 34:2202
100. Krishna R (2009) J Phys Chem C 113:19756
101. Sayeed A, Mitra S, Anil Kumar A, Mukhopadhyay R, Yashonath S, Chaplot S (2003) J Phys Chem B 107:527
102. Shirono K, Endo A, Daiguji H (2005) J Phys Chem B 109:3446
103. Beerdsen E, Dubbeldam D, Smit B (2006) J Phys Chem B 110:22754
104. Ghorai PK, Yashonath S (2004) J Chem Phys 120:5315
105. Du X (2013) J Chem 2013
106. Xiao J, Wei J (1992) Chem Eng Sci 47:1123
107. Bakker WJ, Van Den Broeke LJ, Kapteijn F, Moulijn JA (1997) AIChE J 43:2203
108. Magalhães FD, Laurence RL, Conner WC (1996) AIChE J 42:68
109. Kar S, Chakravarty C (2000) J Phys Chem B 104:709
110. Jobic H, Theodorou DN (2006) J Phys Chem B 110:1964
111. Yashonath S, Ghorai PK (2008) J Phys Chem B 112:665
112. Ghorai PK, Yashonath S, Demontis P, Suffritti GB (2003) J Am Chem Soc 125:7116
113. Yashonath S, Santikary P (1994) J Phys Chem 98:6368
114. Leroy F, Jobic H (2005) Chem Phys Lett 406:375
115. Perez-Ramirez J, Christensen CH, Egeblad K, Christensen CH, Groen JC (2008) Chem Soc Rev 37:2530
116. Cavalcante CL Jr, Silva NM, Souza-Aguiar EF, Sobrinho EV (2003) Adsorption 9:205
117. Gunadi A, Brandani S (2006) Microporous Mesoporous Mater 90:278
118. Liu Z, Fan W, Xue Z, Ma J, Li R (2013) Adsorption 19:201
119. Bonilla MR, Valiullin R, Kärger JR, Bhatia SK (2014) J Phys Chem C 118:14355
120. Coasne B, Hung FR, Pellenq RJ-M, Siperstein FR, Gubbins KE (2006) Langmuir 22:194
121. Bhattacharya S, Coasne B, Hung FR, Gubbins KE (2008) Langmuir 25:5802
122. Crabtree JC, Molinari M, Parker SC, Purton JA (2013) J Phys Chem C 117:21778
123. Chae K, Shi Y, Huang L (2013) J Mater Chem A 1:3886
124. Coasne B, Galarneau A, Gerardin C, Fajula FO, Villemot FO (2013) Langmuir 29:7864
125. Zheng H, Zhao L, Ji J, Gao J, Xu C, Luck F (2015) ACS Appl Mater Interfaces 7:10190
126. García-Martínez J, Li K, Krishnaiah G (2012) Chem Commun 48:11841
127. Yang X-Y, Leonard A, Lemaire A, Tian G, Su B-L (2011) Chem Commun 47:2763
128. Sazama P, Sobalik Z, Dedecek J, Jakubec I, Parvulescu V, Bastl Z, Rathousky J, Jirglova H (2013) Angew Chem Int Ed 52:2038
129. Houžvička J, Jacobsen C, Schmidt I (2001) Stud Surf Sci Catal 135:158
130. Schmidt I, Krogh A, Wienberg K, Carlsson A, Brorson M, Jacobsen CJ (2000) Chem Commun 2157
131. Li Y, Guo W, Fan W, Yuan S, Li J, Wang J, Jiao H, Tatsumi T (2011) J Mol Catal A Chem 338:24
132. Sastre G, Katada N, Suzuki K, Niwa M (2008) J Phys Chem C 112:19293
133. Wang N, Zhang M, Yu Y (2013) Microporous Mesoporous Mater 169:47
134. Katada N, Suzuki K, Noda T, Sastre G, Niwa M (2009) J Phys Chem C 113:19208
135. Kotrel S, Lunsford JH, Knözinger H (2001) J Phys Chem B 105:3917
136. Daniell W, Topsøe NY, Knözinger H (2001) Langmuir 17:6233
137. Simperler A, Bell RG, Foster MD, Gray AE, Lewis DW, Anderson MW (2004) J Phys Chem B 108:7152
138. Sierka M, Eichler U, Datka J, Sauer J (1998) J Phys Chem B 102:6397
139. Pine LA, Maher PJ, Wachter WA (1984) J Catal 85:466

140. Zhou D, He N, Wang Y, Yang G, Liu X, Bao X (2005) J Mol Struct (THEOCHEM) 756:39
141. Zhou D, Bao Y, Yang M, He N, Yang G (2006) J Mol Catal A Chem 244:11
142. Zheng A, Chen L, Yang J, Zhang M, Su Y, Yue Y, Ye C, Deng F (2005) J Phys Chem B 109:24273
143. Chu CTW, Chang CD (1985) J Phys Chem 89:1569
144. Wang Y, Zhou D, Yang G, Miao S, Liu X, Bao X (2004) J Phys Chem A 108:6730
145. Chatterjee A, Iwasaki T, Ebina T, Miyamoto A (1998) Microporous Mesoporous Mater 21:421
146. Yuan SP, Wang JG, Li YW, Jiao H (2002) J Phys Chem A 106:8167
147. Wang Y, Yang G, Zhou D, Bao X (2004) J Phys Chem B 108:18228
148. Meeprasert J, Jungsuttiwong S, Namuangruk S (2013) Microporous Mesoporous Mater 175:99
149. Yang G, Zhou L, Han X (2012) J Mol Catal A Chem 363–364:371
150. Babich IV, Moulijn JA (2003) Fuel 82:607
151. Delannay F (1985) Appl Catal 16:135
152. Nortier P, Fourre P, Saad ABM, Saur O, Lavalley JC (1990) Appl Catal 61:141
153. Alstrup I, Chorkendorff I, Candia R, Clausen BS, Topsøe H (1982) J Catal 77:397
154. Kasztelan S, Grimblot J, Bonnelle JP, Payen E, Toulhoat H, Jacquin Y (1983) Appl Catal 7:91
155. Topsøe N-Y, Topsøe H (1983) J Catal 84:386
156. Maugé F, Duchet JC, Lavalley JC, Houssenbay S, Payen E, Grimblot J, Kasztelan S (1991) Catal Today 10:561
157. Calais C, Matsubayashi N, Geantet C, Yoshimura Y, Shimada H, Nishijima A, Lacroix M, Breysse M (1998) J Catal 174:130
158. Bouwens SMAM, Koningsberger DC, De Beer VHJ, Louwers SPA, Prins R (1990) Catal Lett 5:273
159. Delley B (2000) J Chem Phys 113:7756
160. Morin C, Eichler A, Hirschl R, Sautet P, Hafner J (2003) Surf Sci 540:474
161. Fan J, Wang G, Sun Y, Xu C, Zhou H, Zhou G, Gao J (2010) Ind Eng Chem Res 49:8450
162. Raybaud P, Hafner J, Kresse G, Kasztelan S, Toulhoat H (2000) J Catal 189:129
163. Schweiger H, Raybaud P, Kresse G, Toulhoat H (2002) J Catal 207:76
164. Schweiger H, Raybaud P, Toulhoat H (2002) J Catal 212:33
165. Raybaud P (2007) Appl Catal Gen 322:76
166. Paul J-F, Payen E (2003) J Phys Chem B 107:4057
167. Todorova T, Alexiev V, Prins R, Weber T (2004) Phys Chem Chem Phys 6:3023
168. Byskov LS, Nørskov JK, Clausen BS, Topsøe H (1999) J Catal 187:109
169. Raybaud P, Hafner J, Kresse G, Toulhoat H (1998) Phys Rev Lett 80:1481
170. Cristol S, Paul J-F, Payen E, Bougeard D, Hutschka F, Clémendot S (2004) J Catal 224:138
171. Yao X-Q, Li Y-W, Jiao H (2005) J Mol Struct (THEOCHEM) 726:67
172. Yao X-Q, Li Y-W, Jiao H (2005) J Mol Struct (THEOCHEM) 726:81
173. Todorova T, Prins R, Weber T (2007) J Catal 246:109
174. Todorova T, Prins R, Weber T (2005) J Catal 236:190
175. Harris S, Chianelli RR (1986) J Catal 98:17
176. Chianelli RR, Pecoraro TA, Halbert TR, Pan WH, Stiefel EI (1984) J Catal 86:226
177. Orita H, Uchida K, Itoh N (2004) Appl Catal Gen 258:115
178. Salmeron M, Somorjai GA, Wold A, Chianelli R, Liang KS (1982) Chem Phys Lett 90:105
179. Raybaud P, Hafner J, Kresse G, Kasztelan S, Toulhoat H (2000) J Catal 190:128
180. Sun M, Nelson AE, Adjaye J (2004) J Catal 226:32
181. Topsøe H, Clausen BS, Candia R, Wivel C, Mørup S (1981) J Catal 68:433
182. Chianelli RR, Ruppert AF, Behal SK, Kear BH, Wold A, Kershaw R (1985) J Catal 92:56
183. Okamoto Y, Kubota T (2003) Catal Today 86:31
184. Okamoto Y, Kawano M, Kawabata T, Kubota T, Hiromitsu I (2004) J Phys Chem B 109:288

185. Shido T, Prins R (1998) J Phys Chem B 102:8426
186. Niemann W, Clausen B, Topsøe H (1990) Catal Lett 4:355
187. Louwers SPA, Prins R (1992) J Catal 133:94
188. Clausen BS, Topsoe H, Candia R, Villadsen J, Lengeler B, Als-Nielsen J, Christensen F (1981) J Phys Chem 85:3868
189. Clausen B, TopsØe H (1989) Hyperfine Interact 47–48:203
190. Lauritsen JV, Helveg S, Lægsgaard E, Stensgaard I, Clausen BS, Topsøe H, Besenbacher F (2001) J Catal 197:1
191. Breysse M, Portefaix JL, Vrinat M (1991) Catal Today 10:489
192. Breysse M, Afanasiev P, Geantet C, Vrinat M (2003) Catal Today 86:5
193. Dzwigaj S, Louis C, Breysse M, Cattenot M, Bellière V, Geantet C, Vrinat M, Blanchard P, Payen E, Inoue S, Kudo H, Yoshimura Y (2003) Appl Catal Environ 41:181
194. Song C (2003) Catal Today 86:211
195. Song C, Ma X (2003) Appl Catal Environ 41:207
196. Bezverkhyy I, Ryzhikov A, Gadacz G, Bellat J-P (2008) Catal Today 130:199
197. Ma X, Velu S, Kim JH, Song C (2005) Appl Catal Environ 56:137
198. Ma X, Sprague M, Song C (2005) Ind Eng Chem Res 44:5768
199. Khare GP, Engelbart DR, Cass BW (1999) US Patent 5,914,292
200. Khare GP (2001) US Patent 6,184,176
201. Khare GP, Engelbart DR (2002) US Patent 6,350,422
202. Khare GP, Engelbart DR, Cass BW (2000) US Patent 6,056,871
203. Tawara K, Nishimura T, Iwanami H, Nishimoto T, Hasuike T (2001) Ind Eng Chem Res 40:2367
204. Siriwardane RV, Gardner T, Poston JA, Fisher EP, Miltz A (2000) Ind Eng Chem Res 39:1106
205. Zhang J, Liu Y, Tian S, Chai Y, Liu C (2010) J Nat Gas Chem 19:327
206. Ryzhikov A, Bezverkhyy I, Bellat J-P (2008) Appl Catal Environ 84:766
207. Huang L, Wang G, Qin Z, Dong M, Du M, Ge H, Li X, Zhao Y, Zhang J, Hu T, Wang J (2011) Appl Catal Environ 106:26
208. Huang L, Wang G, Qin Z, Du M, Dong M, Ge H, Wu Z, Zhao Y, Ma C, Hu T, Wang J (2010) Catal Commun 11:592
209. Meng X, Huang H, Shi L (2013) Ind Eng Chem Res 52:6092
210. Greenwood GJ, Kidd D, Reed L (2000) NPRA 2000 annual meeting, AM00-12, San-Antonio, Texas (March 26–28, 2000), 7p
211. Hernández-Maldonado AJ, Yang RT (2004) J Am Chem Soc 126:992
212. Hernández-Maldonado AJ, Yang RT (2002) Ind Eng Chem Res 42:123
213. Liu D, Gui J, Sun Z (2008) J Mol Catal A Chem 291:17
214. Jayaraman A, Yang FH, Yang RT (2006) Energy Fuels 20:909
215. Cychosz KA, Wong-Foy AG, Matzger AJ (2009) J Am Chem Soc 131:14538
216. Marín-Rosas C, Ramírez-Verduzco LF, Murrieta-Guevara FR, Hernández-Tapia G, Rodríguez-Otal LM (2010) Ind Eng Chem Res 49:4372
217. Peralta D, Chaplais G, Simon-Masseron A, Barthelet K, Pirngruber GD (2012) Energy Fuels 26:4953
218. Orita H, Itoh N (2004) Surf Sci 550:177
219. Cheng P, Zhang S, Wang P, Huang S, Tian H (2013) Comput Theor Chem 1020:136
220. Liu P, Lightstone JM, Patterson MJ, Rodriguez JA, Muckerman JT, White MG (2006) J Phys Chem B 110:7449
221. Rodriguez JA, Liu P, Takahashi Y, Nakamura K, Viñes F, Illas F (2009) J Am Chem Soc 131:8595
222. Rodriguez JA (1997) J Phys Chem B 101:7524
223. Henkelman G, Arnaldsson A, Jónsson H (2006) Comput Mater Sci 36:354
224. Bader RFW (1991) Chem Rev 91:893
225. Mittendorfer F, Hafner J (2001) Surf Sci 492:27

226. Mittendorfer F, Hafner J (2003) J Catal 214:234
227. Zhang S, Zhang Y, Huang S, Wang P, Tian H (2012) Appl Surf Sci 258:10148
228. Tominaga H, Nagai M (2010) Appl Catal Gen 389:195
229. Laredo GC, Leyva S, Alvarez R, Mares MT, Castillo J, Cano JL (2002) Fuel 81:1341
230. Will K, Iva S, John A, Alan EN, Murray RG (2004) Energy Fuels 18:539
231. Deena F, Dalai AK, Adjaye J (2003) Energy Fuels 17:164
232. Seunghan S, Kinya S, Isao M (2000) Energy Fuels 14:539
233. Sun M, Nelson AE, Adjaye J (2004) J Mol Catal A Chem 222:243
234. Sun M, Nelson AE, Adjaye J (2005) Catal Today 109:49
235. Edward F (2003) Appl Catal Gen 240:1
236. Piskorz W, Adamski G, Kotarba A, Sojka Z, Sayag C, Djéga-Mariadassou G (2007) Catal Today 119:39
237. Wang P, Wang D, Xu C, Gao J (2007) Appl Catal Gen 332:22
238. Wang P, Wang D, Xu C, Liu J, Gao J (2007) Catal Today 125:263
239. Costa BOD, Querini CA (2010) Appl Catal Gen 385:144
240. Shen W, Gu Y, Xu H, Dubé D, Kaliaguine S (2010) Appl Catal Gen 377:1
241. Morales G, van Grieken R, Martín A, Martínez F (2010) Chem Eng J 161:388
242. Baronetti G, Thomas H, Querini CA (2001) Appl Catal Gen 217:131
243. Hommeltoft SI (2001) Appl Catal Gen 221:421
244. Wang R, Li Y, Guo B, Sun H (2012) Ind Eng Chem Res 51:6320
245. Janik MJ, Davis RJ, Neurock M (2006) Catal Today 116:90
246. Kumar P, Vermeiren W, Dath J-P, Hoelderich WF (2006) Appl Catal Gen 304:131
247. Tang S, Scurto AM, Subramaniam B (2009) J Catal 268:243
248. Cui P, Zhao G, Ren H, Huang J, Zhang S (2013) Catal Today 200:30
249. Berenblyum AS, Katsman EA, Karasev YZ (2006) Appl Catal Gen 315:128
250. Huang C, Liu Z, Shi Q, Xu C, Liu Y (2003) J Fuel Chem Technol 31:462
251. Huang C, Liu Z, Shi Q (2003) J Univ Pet China 27:120
252. Liu Y, Liu Z, Xu C, Zhang R (2005) Chem Ind Eng Prog 24:656
253. Xia RA, Zhang R, Meng X, Liu Z, Meng J, Xu C (2011) Pet Sci 8:219
254. Zhang R, Meng X, Liu Z, Meng J, Xu C (2008) Ind Eng Chem Res 47:8205
255. Qi L, Meng X, Zhang R, Liu H, Xu C, Liu Z, Klusener PAA (2015) Chem Eng J 268:116
256. Ma H, Zhang R, Meng X, Liu Z, Liu H, Xu C, Chen R, Kusener PAA, de With J (2014) Energy Fuels 28:5389
257. Liu Z, Meng X, Zhang R, Xu C, Dong H, Hu Y (2014) AIChE J 60:2244
258. Cui J, de With J, Klusener PAA, Su X, Meng X, Zhang R, Liu Z, Xu C, Liu H (2014) J Catal 320:26
259. Zhou J, Liu X, Zhang S, Mao J, Guo X (2010) Catal Today 149:232
260. Kimura T (2003) Catal Today 81:57
261. Løften T, Blekkan EA (2006) Appl Catal Gen 299:250
262. Smit B, Maesen TL (2008) Nature 451:671
263. Avelino C, Hermenegildo GA (2003) Chem Rev 103:4307
264. Föttinger K, Kinger G, Vinek H (2003) Appl Catal Gen 249:205
265. Torsten MA, Bettina K-C (1999) J Catal 187:202
266. Essayem N, Ben Taârit Y, Feche C, Gayraud PY, Sapaly G, Naccache C (2003) J Catal 219:97
267. Höchtl M, Jentys A, Vinek H (2001) Appl Catal Gen 207:397
268. Chu HY, Rosynek MP, Lunsford JH (1998) J Catal 178:352
269. Asensi MA, Corma A, Martìnez A (1996) J Catal 158:561
270. Demuth T (2003) J Catal 214:68
271. Guo Y-H, Pu M, Liu L-Y, Li H-F, Chen B-H (2008) Comput Mater Sci 42:179
272. Guo Y-H, Pu M, Li H-F, Liu L-Y, Chen B-H (2007) Mater Chem Phys 106:394
273. Hong Li Z, Jun Gong Y, Pu M, Wu D, Han Sun Y, Zhong Dong B et al (2003) Determination of SiO_2 colloid core size by Saxs. J Mater Sci Lett 22(1):33–35

274. Funamoto T, Nakagawa T, Segawa K (2005) Appl Catal Gen 286:79
275. Tomonori K, Takashi A, Shigeru Y (2002) J Mol Catal A Chem 177:289
276. Dupont C, Lemeur R, Daudin A, Raybaud P (2011) J Catal 279:276
277. Badawi M, Cristol S, Paul J-F, Payen E (2009) C R Chim 12:754

Index

A
Acetonitrile, 73, 77
Acid–base interactions, 8
Acridine, 161
Adsorption, 121
Aggregates, 19
Aggregation, 1, 10, 15
Ag–Y zeolite, 157
Alcohols, 77–79, 169
Aldehydes, 77, 85, 169
Alkenes, 17, 77, 136, 164
Alkoxides, 168
Alkylation, 86, 163
Alkylbenzothiophenes, 82
Alkyldibenzothiophenes, 82
Alkylfluorenes, 82
Alkylphenols, 9
Alkylporphyrins, 40, 42, 46, 62
Alkylthiolanes, 82
Alkylthiophenes, 82
ARM model equation, 114
ARM reaction network analysis, 112
Ascarite, 79
Asphaltene, 1, 71, 80
    molecular interactions, 8
Association model, 1
Athabasca asphaltenes, 16, 52, 80–82
Atmospheric equivalent boiling point (AEBP), 128

B
Benzenepolycarboxylic acids, 74
Benzothiophene (BT), 154
B–L method, 130
Born–Oppenheimer approximation, 127
Brønsted acids, 8, 140, 149, 169
Butane, 61, 169
Butanedioic acid, 84
Butene, 163, 168

C
Carbazoles, 160
Carbonate Triangle bitumen, 81
Carbon–sulfur bonds, 113
Carboxylic acids, 8
Catalysts, 2, 39, 84, 121, 135, 163, 169
    bifunctional, 168
    molybdenum, 161
Chlorophyll, 47
Coarse-grained level, 25
Cobalt, 155
Coking, 112
Composite ILs (CILs), 166
Composition modeling, 93
Copper, 129, 165
m-Cresol, alkylation, 165
Critical aggregation concentration (CAC), 15
Critical clustering concentration (CCC), 17
Critical micelle concentration (CMC), 15
Critical nanoaggregate concentration (CNAC), 17
Cu(I)–Y zeolite, 157
Cycloheptane, 76
Cyclohexane, 166
1-Cyclohexyl-2-iminobenzene, 163

Cyclohexylbenzene, 163
Cyclohexylcyclohexene, 163

D
Daqing, 83, 131, 134
Decarboxylation, 47, 48, 81, 115
Demetallization, 39, 50, 61
Density functional theory (DFT), 24
Deoxophylloerythroetio porphyrins (DPEP), 41, 60
Deposition, 15
Desulfurization, 41, 62, 83, 133, 152, 166
α,ω-Di-alkanoic acids, 78
Dibenzothiophene (DBT), 154
Dicarboxylic acid, 74
Dicyclic-deoxophylloerythroetio porphyrins (Di-DPEP), 41
Diffusion, 121
Dimethylcyclopropane, 167
Dimethyldibenzothiophene (DMDBT), 154
Dimethylhexane, 163
Direct desulfurization (DDS), 152
Disulfides, 152
Double layer gum (DG), 133

E
Eley–Rideal reaction pathways, 161
Equation of state (EoS), 1
Etioporphyrins (ETIO), 41, 60
EXAFS, 39

F
Face-to-face stacking, π–π stacking, 9
Faujasite, 138–141
Fluid catalytic cracking (FCC), 121
Fourier transform ion cyclotron resonance mass spectrometry (FT-ICR MS), 39
Furan, 169

G
Gasoline, 4, 95, 112, 121, 140, 156, 163, 166
Gray's model, 12
Gudao, 82, 83, 131

H
Heavy oil, 1, 8, 71, 61, 93, 96
   reaction model, 112
Heptane, 2, 16, 18, 22, 24, 45, 57
   insolubles (C7I fraction), 15
Hexane, 45, 139

Hexene, 168
Hydrodemetallation (HDM), 46, 62, 83
Hydrodenitrogenation (HDN), 83, 133, 160, 163
Hydrodeoxygenation (HDO), 169
Hydrodesulfurization (HDS), 41, 83, 152, 160
Hydrogenation, 39, 61, 132, 152, 159, 169
Hydrogen bonding, 9
Hydrogenolysis, 62, 154, 159

I
Indole, 161
Ionic liquids (IL), 164
Iron, 165
Irreducible small molecules (IM), 113
Isobutane, 163–166
   alkylation, 165
Isobutene, 61, 164
Isomerization, 166

K
Kerogen, 54, 60, 73, 85
Kerosene, 27, 95, 112
Kinetic modeling, 93

L
Langmuir–Hinshelwood mechanism, 161
LCO, 112
LFER, 115
Lloydminster heavy oil, 81

M
Main methyl method (MMM), 104
Maltene, 2, 7, 15, 19, 83, 144
Mercaptans, 152
Metalloporphyrins, 9
4-Methylbenzothiophene (MBT), 154
Micellization, 15
Molecular interactions, 8
Molecular modeling, 121
Molecular simulation, 1
Molecular weight (MW), 4, 80, 97, 117, 128
Molybdenum nitride, 163
Molybdenum sulfide, 152
Monomers, 18
Monte Carlo method (MC), 123

N
N4VO, 41
Nanoaggregation, 16
Naphtha, 134, 152

Index 181

Naphthalene, 76, 149
Naphthenic ring aromatization, 115
Naphthenic ring opening, 115
NaY, 138–143
Nickel, 9, 39, 42, 131, 155, 160
Nickel boride, 82
Nickel porphyrins, 10, 39, 52
NiMoS, 160
Nitrogen, 9, 39, 51, 61, 84, 129, 160
  removal, 39, 84
NMR, 7, 17, 21, 81, 97, 110, 130–133, 149
Non-porphyrin, 39
NOx, 160

O

Octahydrocarbazole, 162
Oil compatibility, 26
Oil deposition, 1
Organonitrogen compounds, 160
Oxovanadium, 9, 46

P

PAC model, 14
Palladium, 168
Particle aggregation control (PAC), 14
Peace River bitumen, 81
Peace River steam-produced bitumen, 81
Pentane, 2, 15, 22, 168
  insolubles (C5I fraction), 15
  isomerization, 166
1,5-Pentanedioic acid, 84
Pentene, 168
Perhydrocarbazole, 163
Perturbed-chain SAFT (PC-SAFT), 28
Petroleum, heavy, 1, 71, 93, 121
Phosphotungstic acid, 164
Pipeline fouling/plugging, 2
Platinum, 166–169
Polycyclic aromatic hydrocarbon (PAH), 12, 16, 133
Pore structure, 121
Probability density functions (pdfs), 104
Pyridine, 161
Pyrrole, 161

Q

QM/MM, 25
Quinoline, 160

R

Reactive adsorption desulfurization (RADS), 156
  removal, 39, 84

Renqiu, 131
Resid, 93
Residue fluid catalytic cracking (RFCC), 62
Rhodo-deoxophylloerythroetio porphyrins (rhodo-DPEP), 41, 42
Rhodo-dicyclic-deoxophylloerythroetio (rhodo-Di-DPEP), 42
Rhodo-etioporphyrins (rhodo-ETIO), 41
Ruthenium ion-catalyzed oxidation (RICO), 7, 71

S

Self-association, 15
Shengli, 83, 131
  asphaltenes, 83
Single layer gum (SG), 133
Solvent extraction, 44
Statistical associating fluid theory equation of state (SAFT EoS), 28
Structure-oriented lumping (SOL), 95, 98
Sulfated zirconia (SZ), 168
Sulfhydryl, 169
Sulfide ring aromatization, 115
Sulfide ring opening, 113, 115
Sulfides, 73, 77, 102, 152, 160
Sulfones, 77
Sulfur, 19, 39, 43, 61, 84, 102, 129
Supercritical fluid extraction and fractionation (SFEF), 61

T

Tetrahydrocarbazoles, 160, 162
Thermal cracking, 113
Thermodynamic models, 26
Thiophene, 152
Transition metals (TMCs), 161
Trimethylpentanes (TMPs), 163
Triple layer gum (TG), 133

V

Vacuum residue (VR), 15, 43, 50, 57, 80, 83, 95, 131, 134
Vacuum residue desulfurization (VRDS), 62
Vanadium, 9, 42, 129
Vanadyl porphyrin, 39
Vapor pressure osmosis (VPO), 4, 20, 82, 97
VGO, 4, 95, 131

W

Wellbore fouling/plugging, 2

X
XANES, 39

Y
Yarranton's model, 13
Yen model, 11
Yen–Mullins model, 11, 23

Z
Zeolite, 121, 131, 135–168
Zinc, 165
Zirconia, sulfated (SZ), 168
ZnO, 156
ZnS, 157
ZSM-22, 168